D1246790

The
Discourses
of Science

Discourses of Science

Marcello Pera

Marcello Pera is professor of the philosophy of science at the University of Pisa.

The University of Chicago Press, Chicago 60637
The University of Chicago Press, Ltd., London

Printed in the United States of America
03 02 01 00 99 98 97 96 95 94   1 2 3 4 5

ISBN: 0-226-65617-9 (cloth)

Originally published as *Scienza e retorica*
©1991, Gius. Laterza & Figli

Library of Congress Cataloging-in-Publication Data
Pera, Marcello, 1943–
[Scienza e retorica. English]
The discourses of science / Marcello Pera ; translated by Clarissa Botsford.
p. cm.
Translation of: Scienza e retorica.
Includes bibliographical references and index.
1. Science—Philosophy. 2. Science—Methodology.
3. Communication in science. I. Title.
Q175.P3826313 1994
500—dc20   94-14169

The paper used in this publication meets the minimum standard requirements for the American National Standard for Information Sciences—Permanence of Paper for Printed Library Materials, ANSI Z39.48-1984.

Contents

You build your plots with logic. Everything takes place
as in a game of chess: here the delinquent, there the victim,
here the accomplice, down there the profiteer. Once the detective
knows the rules and plays the game, the criminal is as good as caught,
and the cause of justice furthered. This fiction drives me crazy.
With logic you can only partially get to the truth.

F. Dürrenmatt, *Das Versprechen*

Preface

But, alas, just when we were going to put all these objections and all the replies to the objections together to get them into some sort of order, why, heavens, they came to a book in themselves.

A. Manzoni, *The Betrothed*

The underlying idea of this book stems from my now more than ten-years-old *Apologia del metodo* (1982). To the few who will exclaim *quantum mutata ab illo!*, I reply, in the same language, *tempora mutantur et nos mutamur cum illis.*

At the time I was firmly convinced that the turn made by the "new philosophy of science" imposes not partial and minor adjustments but a complete reconsideration of the received view of science. For this aim I suggested that if one wants to save certain typical properties of science (such as objectivity, rationality, progressiveness, and so on) without rejecting what authentically new philosophy has brought to the fore (for example, its historical and cultural dimension, incommensurability between theories, meaning variance), one has to transfer science from "the kingdom of demonstration to the domain of argumentation." I still believed, however, that the idea of method, although profoundly revised, was still indispensable.

On this point (not on several others) I have now changed my mind. Further attempts and, above all, extensive historical research (which led me to write *The Ambiguous Frog: The Galvani-Volta Controversy on Animal Electricity*, Princeton University Press, 1991) have convinced me that the idea of scientific method (at least in the version that dates back to Descartes and continues as far as Popper and beyond) is of little or no utility. I thus took to the road of argumentation again with more confidence.

"Argumentation" was still a prudent notion. Later on I made use of "rhetoric" but with considerable embarrassment and unease for three reasons: the concept still had a heavy negative burden even after Ch. Perelman had made his laudable efforts to vindicate it; Perelman himself had been reluctant or opposed to its extension to scientific contexts; and when it eventually became widespread, it also became loaded with connotations that are not related, or are contrary, to my project. The point is that much of the work, even good work, done under the label of "science and rhetoric," "rhetoric in science," and the like, is mainly concerned with sociological, hermeneutic, or communication questions regarding scientific texts, their making, presentation, and diffusion. Either that, or it aims at showing that facts are words.

This is not my view; it could not be further from it. I do not believe, say, that the sense of a molecule "is an effect only of words, numbers, and pictures judiciously used with persuasive intent," as A. Gross writes in his *The Rhetoric of Science* (Harvard University Press, 1990). I consider B. Latour and S. Woolgar's view (*Laboratory Life,* Princeton University Press, 1979), that "facts are socially constructed," as true if trivial, and false if interesting. Nor do I aim at unmasking "The Rhetoric of the Radical Rhetoric of Science," as a paper by J. E. McGuire and T. Melia suggests. I wish, rather, to focus on and understand scientific *discourse* and, through it, the *value* of science as a cognitive endeavor.

As for the discourse, I shall make a distinction. I shall refer to "rhetoric" as the *practice* of persuasive argumentation (or "the *act* of persuading," which is one of its meanings registered by Melia), and to "dialectics" as the *logic* of such a practice or act. I thus agree with L. J. Prelli (*A Rhetoric of Science,* University of South Carolina Press, 1989), who defines rhetoric as "the use of symbols [I would say of arguments] to induce cooperative acts and attitudes," and I accept the distinction introduced by Aristotle and recently restored by J. D. Moss in her *Novelties in the Heavens: Rhetoric and Science in the Copernican Controversy* (University of Chicago Press 1993), between dialectics, which makes use of *logos* and rhetoric, and which, in addition, employs *ethos* and *pathos.* But unlike Moss, and probably like Aristotle (though his view is here controversial), I do not think that in scientific contexts the former can be separated from the latter. In my view, scientific discourse is not rhetorical in an ornamental way, as if scientific claims could be proved on certain grounds (for example, Galileo's "sensory experiences and mathematical demonstrations" or Newton's "phenomena") and made appealing or palatable on others (for example, Galileo's dialogical form or

Newton's mathematical arrangement). Scientific discourse is rhetorical in a constitutive way, because scientific claims are accepted only if they persuade the audience (community) within which they are put forward and debated through an exchange of arguments and counterarguments whose outcome depends in no way on the passions and personal commitments of the protagonists, or on the style of the text. To outline the logic of debate (dialectics),with its factors, techniques, and strategies, is one of my aims here.

As for the cognitive value of science, just because rhetoric, in my sense, neglects *ethos* and *pathos* and concentrates on *logos,* does not imply that it is deceptive or weak; simply that it changes its source and way of legitimation. In the standard methodological model, scientific research is a game with two players: the scientist's inquiring mind that asks questions, on the one hand; nature that provides answers by saying "yes" or "no," on the other. An impartial arbiter—method—is in the middle, ascertaining whether the game is conducted well and when it is over. As it is guided or forced (or "mastered" as Bacon put it) by the rules of the arbiter, nature speaks out, and "knowing" amounts to the scientist's *recording* of nature's true voice, or *mirroring* its real structure. In what I call the "dialectical model," there are three players: an individual or a group of individuals, nature, and another group of individuals that debates with the first according to the factors of scientific dialectics. Here nature reacts to a cross-examination, there is no impartial arbiter, and "knowing" amounts to the community's *agreeing upon* nature's correct answer. This does not replace objectivity with "solidarity" or rationality with "routine conversation" (as Rorty puts it), because agreement among the members of a community is not merely conversational; it is constrained, although not imposed or dictated, by nature. Providing an appropriate epistemological framework for this constraint is another aim of my work.

I am painfully aware that a project of this kind requires competence in different fields, from general philosophy to logic (including the much neglected informal logic), from philosophy of language and epistemology to the history and philosophy of science. I have tried to carry out this project as far as I have been able. I have two main desires: to suggest that my project, defective as it may be, is worth pursuing, and to show that approaching science from the point of view of its discourses may shed more light on its cognitive properties, today so frivolously mocked because yesterday those properties were so deeply misunderstood. For the former aim, I cannot but submit myself to the judgment of the interested reader; for the latter, I must say that much of the work in the current philosophy of science makes me feel that this approach is already at work, although it still lacks a well-defined framework.

I started thinking and working on this book in 1984, in one of the most intellectually exciting environments I have ever experienced: the Center for Philosophy of Science at the University of Pittsburgh. After the Italian edition came out (of which the present is a highly revised version) I did much of the work for the English edition in 1991 when I was a visiting fellow at the Department of Linguistics and Philosophy at MIT. I have much benefited from discussion with Thomas Kuhn, Peter Hempel, Adolf Grünbaum, Wes and Merrilee Salmon, I. Bernard Cohen, and Nick Rescher, whose book *Dialectics* I take as an excellent pioneering contribution to the field. I am very grateful to Kuhn for many friendly private discussions, and I am indebted to him for some of his fundamental philosophical hints that we both consider to have been misunderstood. This book intends to follow a course he has traced, although he has not yet systematically developed it.

I have debts of gratitude with my fellow in philosophy and wine, Peter Machamer, whose penetrating remarks and comments, in different occasions over many years, have showed me certain points that needed, and probably still need, to be made more clear; with Rachel and Larry Laudan, whose ideas have always been stimulating for me even when I decided to disappoint them by retiring from the methodological business; with Tom Nickles, whose friendly confidence in my views has incomprehensibly always been higher than mine; with Aristides Baltas, for the many night and day conversations we have spent on several aspects of this book, and for his optimism (Greeks sometimes leave aside their sense of tragedy where their friends are concerned) that the book would eventually come to light; with Paul Horwich, who has definitively corrupted me with his deflationist theory of truth; with Bob Richards, for his helpful remarks on my treatment of Darwin and on other points; with Paul Feyerabend, for encouraging and provoking me in many public and private discussions or, to put it in Galilean terms, "altercations," which is his and my favorite way of engaging in philosophy and other more serious things. On February 11, 1994, when this book was in proof, Paul passed away. He had just finished writing his autobiography, *Killing Time,* and was enjoying some of the best moments of his life. The community of philosophers lost a protagonist who contributed much to challenge the profession. I miss a sweet, lovely friend.

Several other friends have helped me with comments and remarks. It is a pleasure to acknowledge M. Finocchiaro, who, as a reader of the original Italian book and then as an (anonymous) referee for the English manuscript, has been generous in his criticism—now stimulating, now devastating—but always, I am sure, aimed in the right direction. If I have not always followed his suggestions it is at my expense but also to his advantage, because I would have had to transform my book into his. F. Barone, who irresponsibly introduced me to philosophy, has once again taught me something important by putting his finger on some weak points. My friends Gianni Federspil, Maurizio Mamiani, Pierluigi Barrotta, and Alberto Mura have given me the assurance of counting on diligent and severe readers. Clarissa Botsford has carefully translated the manuscript and patiently resisted my attempts to show that, compared to Italian, English is an unnatural language. Readers should be grateful to Bill Shea, who has restored the truth by putting many linguistic expressions in order.

But most of all I warmly thank my students at the Universities of Catania and Pisa. They are the ones who have acted as innocent guinea pigs for my project and who have seen it growing and taking on a definitive form. If it has improved, this is due mainly to their questions, and to my efforts to answer them, though I am aware that many of their objections still have no answers. To avoid despair, I decided to rely on Manzoni, who already explained why one cannot solve all objections without starting the book over again and writing another one. However, as he added: "one book at a time is enough, if indeed it is not too much."

The Cartesian Syndrome

And now, after all methods, so it is believed, have been tried and found wanting, the prevailing mood is that of weariness and complete indifferentism—the mother, in all sciences, of chaos and night, but happily, in this case, the source, or at least the prelude, of their approaching reform and restoration.

I. Kant, *Critique of Pure Reason*

The recent revolution in the philosophy of science has already spawned a highly specialized literature. One crucial point, nonetheless, has had little, if any, attention: how is it that the traditional image of science has been completely overturned? Whereas science was once commonly taken as the only form of rational knowledge, why is it nowadays so often considered neither more nor less rational than any other form of culture?

Since my aim in this Introduction is to set up a conceptual framework of problems around which to string my discourse, rather than merely to relate history, I will tell a philosophical story. The protagonists I have chosen are not specific flesh-and-blood individuals but typical characters in a play—though I believe the ideas, opinions, views, and ways of thinking I ascribe to the latter are not too far from those held in real life by the former. I am well aware that this expressionistic technique does violence to the history of ideas (just as expressionism in art does violence to reality), but I am also convinced that an expressionistic story is more effective and instructive than a detailed history. One can draw a moral from a story; to try to do so from history can be useless though not always detrimental.

I will divide my story into three acts. The transition from the second to the third act marks the crucial point for overturning the traditional image of science, but it is in the first act that the premises are laid down. As in a Greek tragedy or in a play by Pirandello, here too the destiny of each protagonist is determined from the start.

Let us begin then with *Act 1*. Taking into account the qualities the traditional image of science ascribes to science—that it is certain, infallible, universal, and objective—there are good reasons for calling it *science as demonstration*. This image is made up of two components, which must be examined separately although traditionally they are interconnected: the epistemic and the methodological.

The epistemic component states that science is based on certain data through which we acquire knowledge of reality. These data can be taken differently. They can be experimental, like Galileo's "sensory experiences," or intellectual, like Descartes' "clear and distinct ideas." In both cases, they guarantee that scientific knowledge grasps reality, either because reality reveals itself through a process starting with pure perception or because the structure of reality manifests itself through a chain of inferences stemming from the pure principles of the mind. The "dogma of immaculate perception" or its counterpart, the "dogma of immaculate conception," as we may call them, thus constitute the first pillar, the prime element of faith, of science as demonstration.

The methodological component integrates the epistemic component. It states that science provides knowledge by making use of a method which allows us to process data correctly. Like data, method too can be understood in different ways. It can be an *organon* for inferring cognitive conclusions from observations, as in Bacon; a set of *regulae ad directionem ingenii* for solving cognitive problems, as in Descartes; a set of *regulae philosophandi* to establish how to conduct an inquiry, as in Newton; or a *libra* to weigh the degrees of probability of rival hypotheses, as in Leibniz; and so on. In each case, method guarantees that, if the information is correct, so will the conclusion be. The dogma of method is thus the second pillar, or element of faith, of science as demonstration.

As this dogma plays a more important role in our story than the other, let us consider briefly the way it acts and what hopes it offers.

Bacon represented scientific method as a ruler or compass that "goes far to level men's wits, and leaves but little to individual excellence because it performs everything by the surest rules and demonstrations."[1] In this view, method is a tool for achieving truth and putting an end to scientific controversies. Descartes shared this view: "whenever two persons make opposite judgments about the same thing, it is certain that at least one of them is mistaken, and neither, it seems, has knowledge. For if the reasoning of one of them were certain and evident, he would be able to lay it before the other in such a way as eventually to convince his intellect as well."[2] Method is precisely what convinces all intellects, both because it contains "the primary rudiments of human reason"[3] and because it consists of "reliable rules which are easy to apply, and such that if one follows them exactly, one will never take what is false to be true or fruitlessly expend one's moral efforts, but will gradually and constantly increase one's knowledge till one arrives at a true understanding of everything within one's capacity."[4] Leibniz's view was no different. It is well known that he greeted Descartes' rules with great irony: "I almost feel like saying that the Cartesian rules are rather like those of some chemist, or rather: take what is necessary, do as you ought to do, and you will get what you wanted."[5] Leibniz's irony, however, was not directed at Descartes' methodological model in itself but at its technical inadequacy. Like Bacon and Descartes, Leibniz too believed that, thanks to the rules of his calculus, "all truth can be discovered by anybody with a secure method [*methoda certa*], to the extent that, making use of reason, they may be obtained by data available even to the greatest and trained talent, with only the difference of readiness, whose importance is greater in action than in meditation and discovery."[6] And like Bacon and Descartes, Leibniz considered method (his universal calculus) an instrument for putting an end to controversies: "when controversies arise, there is no more need for discussion between two philosophers than there is between two calculators. All the two need to do is to sit down at a table, pen in hand (having called a friend if they wish), and mutually declare: *let us calculate.*"[7]

Act 2 of our story opens with this dream of a universal calculus. Since even the most radical expressionist would not be able to give the whole picture, let us go straight to its conclusion. After a series of impressive successes, the first pillar of science as demonstration, the epistemic component, starts crumbling under the weight of its own construction. The birth of non-Euclidean geometry, the crisis of the foundations of mathematics, the rejection of associationist psychology, as well as great intellectual innovations such as relativity and quantum theories—all this showed that not even the clearest and most distinct concepts (space, time, cause, substance, and number, for example) are beyond revision, and that not even the purest perceptions (say, of figures and movement) are without distortions.

At this point, tragedy is unavoidable, but it does not come about because, while the first pillar falls down, philosophers manage to hold up the whole construction by reinforcing the second pillar. On the one hand, they surrender to the weakening of the epistemic component of science as demonstration—from certainty to truth, from truth to probability, from probability to verisimilitude. On the other hand, though they come to change the position of method—from the "context of discovery" to the "context of justification"—they still stick to a few central ideas which are typical of the methodological component. From this point of view, there is little substantive difference between Bacon, Descartes, Leibniz, Newton, Whewell, Mill, etc., and between Popper, Lakatos, Laudan, etc. Although the rules and aims of their methodologies are markedly different, the underlying sense of their project has stayed the same. More precisely, they have all continued to hold the following theses:

First thesis. There is a universal and precise method that demarcates science from any other intellectual discipline.

Second thesis. The rigorous application of this method guarantees the achievement of the aim of science.

Third thesis. If science possessed no method, it would not be a cognitive and rational endeavor.

Since there is good reason to consider Descartes the eponymous hero of this story, I will call any program of philosophy of science based on these theses a *Cartesian project*; in particular, I will call the third thesis the *Cartesian dilemma,* or I will refer to it as the *Cartesian syndrome.*[8]

Since they have suckled the Cartesian project with the milk of their philosophical training, a few examples are enough to show that many eminent scientists and philosophers are still dependent on it.

As for the first thesis of the project, consider the following claim made by a psychologist:

> [T]o judge whether a given discipline is or is not scientific is possible without value implications; it necessitates nothing but a commonly agreed definition and standard of scientific procedure. Such a definition and such standards exist, and may be found in the writings of logicians and philosophers of scientific methodology; those who are acquainted with these writings will agree that in spite of occasional disagreements on minor issues there is an overwhelming amount of agreement on the main points.[9]

Philosophers think the same. Just as Kant sought a "standard weight and measure to distinguish sound knowledge from shallow talk,"[10] so Popper has looked for scientific method to provide a "clear line of demarcation between science and metaphysical ideas,"[11] and so has Lakatos sought to establish "universal definitions of science," or "universal conditions under which a theory is scientific," or "sharp criteria using which one can compare rival fabrications both in physics and in history."[12]

The attitude of Popper and Lakatos towards method also serves well to illustrate the influence of the second thesis of the Cartesian project. According to Popper, abiding by his own methodological rules allows one to obtain theories ever closer to reality; according to Lakatos, the methodology of scientific research programs provides a tool for evaluating which of two alternative programs is epistemically superior. Hence the view of method as "the discipline of rational appraisal of scientific theories — and of criteria of progress."[13]

As far as the third thesis of the project is concerned, a typical source is again Lakatos. In his view, if there were no clear, impersonal methodological rules, one would have to "abandon efforts to give a rational explanation of the success of science"[14] and interpret this success in psychological or sociological terms only: "Often these rules, or systems of appraisal, also serve as 'theories of scientific rationality', 'demarcation criteria' or 'definitions of science'. Outside the legislative domain of these normative rules there is, of course, an empirical psychology and sociology of discovery."[15]

So we are left with either methodological rules or "mob psychology."[16] Attached as he was to these two horns of the Cartesian dilemma, Lakatos could not conceive of any middle ground — at least any *rational* middle ground — between domination by rules and domination by irrationality.[17] It was thus inevitable that when the first horn crumbled, greater emphasis would be placed on the second.

Here we come to *Act 3* of our story, in which the final collapse of the image of science as demonstration comes to the fore. Once the epistemic component had definitely been knocked down, the "new philosophy of science" started attacking the methodological component too. As a result, the first two theses of the Cartesian project were rejected: the first because "the idea of a method that contains firm, unchanging, and absolutely binding principles for conducting the business of science meets considerable difficulty when confronted with the results of historical research"; the second because historical research itself shows that the greatest scientific progress has always taken place precisely because "some thinkers either decided not to be bound by certain 'obvious' methodological rules, or because they unwittingly broke them."[18]

Nevertheless — and this is the really crucial point in the whole story — while the first two theses of the Cartesian project have been rejected, at least some exponents of the new philosophy of science have conserved the third and transformed it from a counterfactual conditional into an assertoric statement. It is this conservation and transformation that explains why the traditional image of science as the only form of rational knowledge has been turned into the currently fashionable image of science as a form of rational (or irrational) culture like any other. Such an about-face is the effect of a hidden Cartesian syndrome.

Let me take a few examples to give an idea of the persistence of this syndrome.

The historian Vasco Ronchi is a good case to start with. Before Polanyi and Kuhn claimed that the acceptance of a new theory depends on an "experience of conversion," and Feyerabend theorized that theory choice is not just a matter of comparison between intrinsic merits, Ronchi had reached the same conclusion. Discussing Newton's optical theory, he wrote: "It is truly marvelous, one might say inexplicable, that such an incoherent and shaky construction, full of contradictions and lacunae, managed to convince the outright majority of eighteenth century physicists, and spread outside the technical realm, like a great scientific discovery."[19]

Ronchi did not stop there. He went on to claim that, considering the situation of optics in Newton's time, "to accept one theory or another is a matter of taste and opinion. Since there is no perfect theory in the midst of all the imperfect, and therefore equally untrue ones, the choice is either personal or dictated by contingent reasons which have nothing to do with the theories themselves. Much depends on finding a powerful and active upholder: what we would call an 'apostle'."[20]

In more recent times, Ronchi went even further. Bringing a "trial against science," in which he claimed that "scientific ideas are defended with the same propogandistic means as political ideas,"[21] he concluded:

> History teaches us that the most important results, the most extraordinary breakthroughs, are never the result of logic or rationality. Though good luck and chance are often responsible for important discoveries, it is a fact that the most powerful instrument a researcher or scholar has at his disposal is faith in success, the tenacity against the innumerable difficulties and the inertia and incomprehension of his peers who resist anything new, especially when the breakthrough is profound and truly important—in short, his enthusiasm and strong-willed desire to achieve his goal.[22]

Ronchi did not explicitly proclaim that in science "anything goes," because, as he wrote, although it is true that "neither logic nor rationality fire the motor of progress . . . it is equally true that if faith and enthusiasm become allies of logic and rationality, then one can expect even more brilliant results."[23] The way Ronchi views the growth of science, though, is a typical symptom of the Cartesian syndrome: since it is not logic that guides science, the acceptance of scientific theories is purely a "matter of taste" dependent on "contingent reasons that have nothing to do with the theories themselves."

Feyerabend, with a considerably deeper philosophical grounding, is on a similar wavelength. From a historical point of view, one can reasonably say that, with Feyerabend, the long tradition of methodology dies, and one can rightly interpret his *Against Method* as the death certificate of Descartes' *Discourse on Method,* and his "anything goes" as the epitaph carved onto the gravestone of Descartes' "certain and simple rules." What does not die with Feyerabend, however, is the typically Cartesian idea that the only alternative to method is irrationality. Consider this: "It is clear that allegiance to the new ideas will have to be brought about by means other than arguments. It will have to be brought about by irrational means such as propaganda, emotion, ad hoc hypotheses, and appeal to prejudices of all kinds."[24]

Or this:

> And if the old forms of argumentation turn out to be too weak a contrary cause, must they not, then, either give up, or resort to stronger and more "irrational" means? (It is difficult, and perhaps entirely impossible, to combat the effects of brainwashing by argument.) Even the most puritanical rationalist will then be forced to leave argument and use, say, propaganda, not because some of his arguments have ceased to be valid, but because the psychological conditions which enable him to effectively argue in his manner and thereby to influence others have disappeared.[25]

Or even this:

> Now this reference to tests and criticisms which is supposed to guarantee the rationality of science and, perhaps, of our entire life, may be either to well-defined procedures without which a criticism or test cannot be said to have taken place, or it may be purely abstract so that it is left to us to fill it now with this, and now with that concrete content. The first case has just been discussed. In the second case we have but a verbal ornament, just as Lakatos' defense of his own "objective standards" turned out to be a verbal ornament.[26]

Opinions such as these can be justified only on the assumption that, since method and logic do not exist, science can be explained only in psychological or sociological terms—precisely the two horns of the Cartesian dilemma. Feyerabend is so attached to these two horns that he claims the only alternative to the neutral verdicts of method is personal opinion.[27] Thus he can well be considered the last coherent Cartesian, paradoxical as it may seem.[28]

Rorty's is another case in point. He too, arriving as he did after the ground had already been well trodden, rejects the first two theses of the Cartesian project; and he too considers vain the efforts of that "host of philosophers—roughly classifiable as 'positivist'—who have spent the last hundred years trying to use notions like 'objectivity', 'rigor' and 'method' to isolate science from non-science," and invites us to abjure the idea that "following that method will enable us to penetrate beneath the appearances and see nature 'in its own terms'."[29] But when he comes to construction, Rorty replaces the idea of method with the idea of "routine conversation" or "good epistemic manners" or hermeneutics,[30] with the effect that "we shall not think there is or could be an epistemologically pregnant answer to the question 'What did Galileo do right that Aristotle did wrong?' any more than we should expect an answer to the questions 'What did Plato do right that Xenophon did wrong?' or 'What did Mirabeau do right that Louis XVI did wrong?'."[31] What we should say, instead, is that Galileo found "the right jargon"[32] and "the question whether he was 'rational' . . . is out of place,"[33] because "to be rational . . . is to be willing to pick up the jargon of the interlocutor,"[34] or because "science does not *have* a secret of success."[35]

Last but not least come the sociologists. They too have absorbed the results of the new philosophy of science and criticized the first two theses of the Cartesian project, but they too raise the white flag when facing the third thesis. Their "strong program" works on the same premise. What, if not the Cartesian dilemma, warrants the inference that, since science does not possess a universal method, "scientific theories, methods, and acceptable results are social conventions"?[36] Mary Hesse has written that "it is only a short step from this philosophy of science to the suggestion that adoption of such criteria, which can be seen as different for different groups at different periods, should be explicable by social rather than logical factors."[37] But this is far from obvious. It is a short step if one maintains that the only alternative to the logic and methodology of science is social conventions, but it is neither short nor safe if one casts doubt on the dilemma itself.

Thomas Kuhn is absent from this list of neo-Cartesians, but not inadvertently. Kuhn is both an eminent historian and philosopher and a poor caricature, with the former maintaining one thing and the latter saying another. It must be admitted that Kuhn did much to encourage his caricaturists, because in his *Structure of Scientific Revolutions* he presented his readers with one of the very phenomena he was describing, the *Gestalt switch.* As a matter of fact, the book could equally be interpreted as the quacking of an irrationalist duck and the squeaking of a rationalist rabbit—of a species never yet contemplated in manuals of epistemological zoology, with the duck on the even pages and the rabbit on the odd pages. Here is a small anthology of extracts of the sounds of the two incommensurable animals:

Kuhn's Duck	Kuhn's Rabbit
Just because it is a transition between incommensurables, the transition between competing para-	Because scientists are reasonable men, one or another argument will ultimately persuade many of them.

digms cannot be made a step at a time, forced by logic and neutral experience. Like the gestalt switch, it must occur all at once (though not necessarily in an instant) or not at all.[38]

But there is no single argument that can or should persuade them all. Rather than a single group conversion, what occurs is an increasing shift in the distribution of professional allegiances.[39]

The transfer of allegiance from paradigm to paradigm is a conversion experience that cannot be forced.[40]

As in political revolutions, so in paradigm choice—there is no standard higher than the assent of the relevant community. To discover how scientific revolutions are affected, we shall therefore have to examine not only the impact of nature and of logic, but also the techniques of persuasive argumentation effective within the quite special groups that constitute the community of scientists.[41]

Individual scientists embrace a new paradigm for all sorts of reasons and usually for several at once. Some of these reasons . . . lie outside the apparent sphere of science entirely.[42]

We must therefore ask how conversion is induced and how resisted. What sort of answer to that question may we expect? Just because it is asked about techniques of persuasion, or about argument and counterargument in a situation in which there can be no proof, our question is a new one, demanding a sort of study that has not been previously undertaken.[43]

A decision between alternate ways of practicing science is called for. . . . A decision of that kind can only be made on faith.[44]

When paradigms enter, as they must, into a debate about paradigm choice, their role is necessarily circular. . . . Yet, whatever its force, the status of the circular argument is only that of persuasion.[45]

Though the historian can always find men—Priestley for instance—who were unreasonable to resist for as long as they did, he will not find

It makes a great deal of sense to ask which of two actual competing theories fits the facts *better*. Though neither Priestley nor Lavoisier's the-a better point at which resistance becomes illogical or unscientific.[46]

ory, for example, agreed precisely with existing observations, few contemporaries hesitated more than a decade in concluding that Lavoisier's theory provided the better fit of the two.[47]

To do justice to Kuhn, it must be said that he always resisted the duck interpretation and never liked the caricatures with which people depicted him. He was perhaps tempted by the Cartesian syndrome but was little if at all affected by it. If we read carefully, he never actually suggested that the only alternative to method is irrationality. Rather, in the odd (rabbit) pages of the *Structure,* and even more clearly in his later writings,[48] he proposed a *new way* of understanding scientific rationality by replacing the old view based on method with a different one grounded in persuasive argumentation. It must also be admitted that, despite this proposal and his repeated protests against the ongoing interpretation of his view, he has not yet developed a systematic and detailed picture of his new way. This is one reason why he has been superficially considered an irrationalist. The other, more important reason is that the prevailing Cartesian syndrome prevented philosophers from using the right categories needed to comprehend his new notions of rationality and new image of science.[49]

Let me sum up, then, and lay my cards on the table. I have presented the Cartesian project and shown that, although it is still influential among scientists and philosophers, it has been strongly attacked. As an effect of this attack, the *methodological model* of science has been turned into a *counter-methodological model* with at least three variations. The first is anarchist, in which cognitive claims and epistemic evaluations depend on "taste," "reasons which have nothing to do with theories," "propagandistic means" (Ronchi), or "irrational means" and "means other than arguments" (Feyerabend). The second is sociological, in which these claims and evaluations are the effects of "sociological rather than logical factors" (Bloor). The third is post-philosophical, and states that the very issue of epistemic evaluation is "out of place." I have also tried to show, however, that this substitution is not as radical as it seems. The counter-methodological model conserves the core of the old methodological model, that is, the Cartesian dilemma between method and irrationality.

My final aim in this book is not to rescue the methodological model but to find a way out of the Cartesian dilemma. Dialectics, not sociology, psychology, or hermeneutics, will be my candidate to replace method. I will thus take the position suggested but not fully developed by Kuhn, and

I will try to draft and elaborate upon a different image of science that I will call the *dialectical model.*

The position I will try to reach is uncomfortable and demanding. It is uncomfortable because it lies between two extremes, one vulnerable, the other fashionable. It is demanding because, in order to avoid being drawn in by either extreme, one cannot simply stop half-way; an altogether new point of view must be found. If dialectics is not considered mere conversation or verbal decoration, if it is forced to become a constitutive part of science, then new meanings will have to be found for old concepts.

The reason for this can be seen schematically. The methodological model views science as a game between two players: the researcher proposes, and nature—with its ringingly clear "yes" or "no"—disposes. In the counter-methodological model, the situation is the same, the only difference being that nature's voice is so weak that it is drowned out by the researcher's, who ultimately becomes nature's ventriloquist, providing the desired answers. The dialectical model is different; it requires three players: a proposer who asks questions, nature that answers, and a community of competent interlocutors which, after a debate hinging on various factors, comes to an agreement upon what is to be taken as nature's official voice. In this model nature does not speak out alone. It only speaks *within* the debate and *through* the debate.

This changes things radically. If nature's voice is that which issues from a debate, then the typical intrinsic properties of cognitive claims—for example, whether they are true, probable, progressive, acceptable, and so on—have to be redefined in terms of the course and outcome of the debate. This task is particularly complicated because it clashes with an old, classical objection dating back at least to Plato: a debate aims to persuade its participants, but is not the persuasion of participants in a debate different from the intrinsic cognitive value of the claim being discussed? A bridge between these two views is clearly needed.

Although the dialectical model occupies an uncomfortable and demanding position, it also promises considerable advantages. The main one is philosophical. If it were possible to build the missing bridge between persuasion and scientific knowledge, then one would be in the best position for overcoming the tension between "internal" and "external," and between normative and descriptive philosophies of science. A dialectics *proper to* science (and not just a normal conversation valid in any field of culture) should be able to show how external factors become internal and how internal factors are conditioned by external ones.

The other advantage of the dialectical project is cultural. Taken as an ideology (scientism), the image of science as demonstration has given rise to the romantic complaint that knowledge sets us apart from reality and life. The image of science as a "conceptual grid," a "form of culture," or "will to power" now risks producing another ideology (something like the "indifferentism" concerning metaphysics denounced by Kant) and another complaint—that science does not even know what it claims to know. The dialectical model may do better: it may help us put science in a more human light, come to understand its place between culture and nature, and appreciate it for what it can give, without blaming it for what it cannot offer.

This is my aim. In order to achieve it I shall proceed by stages. First (chapters 1 and 2) I shall try to show that the first horn of the Cartesian dilemma is truly untenable and that one can escape from the second by replacing "science as demonstration" with "science as argumentation." Next, I shall examine samples of scientific argumentation (chapter 3) and the logic regulating it (chapter 4). The solution provided by the dialectical model to some of the more delicate problems in the current philosophy of science, such as those concerning objectivity, rationality, and truth (chapter 5), will allow me to deal, finally, with the issues of theory preference (chapter 6) and scientific progress (chapter 7).

One

The Paradox of Scientific Method

Descartes wrote and rewrote his book, *On Method,* many times; and yet, as it is now, it is useless. Whoever perseveres over a period of time in scrupulous inquiry, will have to change method sooner or later.

J. W. Goethe, *Maximen und Reflexionen*

All good scientists, doctors, observers, and thinkers do what Copernicus used to do: they turn data and methods upside down to see if they are any better.

Novalis, *Fragmente*

1.1 Scientific Procedure

It might seem odd that some of the most influential ideas about the role of scientific method are to be found in the middle of a book devoted to love, which—as we all know—is not necessarily blind but certainly far from methodical: Plato's *Phaedrus.* At one point, Socrates is led into talking about rhetoric and, in order to illustrate his point, he makes a comparison with medicine.

> *Socr.* Well, look here: Suppose someone went up to your friend Eryximachus or his father Acumenus, and said, "I know how to apply such treatment to a patient's body as will induce warmth or coolness, as I choose; I can make him vomit, if I see fit, or go to stool, and so on and so forth. And on the strength of this knowledge I claim to be a competent physician, and to make a competent physician of anyone to whom I communicate this knowledge." What do you imagine they would have to say to that?
>
> *Phaedr.* They would ask him, of course, whether he also knew which patients ought to be given the various treatments, and when, and for how long.

> *Socr.* Then what if he said, "Oh, no: but I expect my pupils to manage what you refer to by themselves"?
>
> *Phaedr.* I expect they would say "The man is mad: he thinks he has made himself a doctor by picking up something out of a book, or coming across some common drug or other, without any real knowledge of medicine."[1]

What has this man failed to do? Socrates says he has failed to follow "the way to reflect about the nature of anything," that is, he has not followed a method, and that "to pursue an inquiry without doing so would be like a blind man's progress."[2] Socrates' conclusion is to be stressed:

> *Phaedr.* I certainly think that would be an excellent procedure.
>
> *Socr.* Yes. In fact I can assure you, my friend, that no other scientific method of treating either our present subject or any other will ever be found, whether in the models of the schools or in speeches already delivered."[3]

Some of the most influential ideas at the root of Western philosophical reflection on science—including what we have called the Cartesian project—are to be found in these passages. Plato claims: (a) that science is a well-structured discourse; (b) that method makes science a subject matter that can be taught; and (c) that method can be used as a criterion of demarcation between what is scientific and true, and what is not. In addition to these three theses, the following two must be added: (d) Plato claims that the method of science is universal and unique, and (e) that method consists of a well defined and appropriate succession of steps. Thesis (d) is derived from the fact that a correct method allows one to reason "about the nature of anything," and about "our present subject or any other." Thesis (e) is the result of the fact that scientific discourse is not like a speech in which the various parts are "thrown out at haphazard,"[4] or like the epigram written for Midas the Phrygian in which "it makes no difference what order the lines come in,"[5] or like the speeches of rhetoricians who never respect a given order; rather, it is "constructed like a living creature,"[6] an assemblage of harmonic parts.[7] It is these two theses that concern us here.

Let us start with (d). What is method? This idea contains at least three different *explicanda* which must be distinguished.

1. First, scientific method is a *procedure,* a global strategy that indicates an ordered series of moves (or stages, or steps, or operations) which a scientist must carry out (or run through) in order to achieve the aims of science. This is the sense conveyed by the passages quoted from Plato,[8] as well as by such expressions as "deductive method," "inductive method," "hypothetico-inductive method," and the like.[9]

2. Second, scientific method is a set of *rules,* or norms, or prescriptions which govern each step of the procedure. This is what "method" meant for Bacon[10] and Descartes,[11] and, in our day, for Popper[12] and Lakatos,[13] although, for the first two, rules pertained to discovery and, for the second two, they concern justification.

3. Third, scientific method is a set of either conceptual or material *techniques* for making the moves required by the procedure. This is what is meant when one speaks of methods (or techniques) of observation, classification, calculation, conducting experiments, and so on. It is said, for example, that sociology uses the "method" of sampling, psychology the "method" of spoken thought, psychoanalysis the "method" of free association,[14] and so on.

I shall deal with the first *explicandum* in this section, and with the others in the next two sections. Since we are looking for exact *explicata* we must first establish some requisites. Two in particular are essential:

Adequacy. The *explicata* must save cases recognized as being exemplary of scientific practice.

Precision. The *explicata* must allow unambiguous discrimination between inquiries that satisfy them and those that do not.[15]

Whether science has a method, as Plato declared, and as the first thesis of the Cartesian project maintains, or whether the idea of scientific method is "unrealistic," "pernicious," or "detrimental to science," as Feyerabend has claimed,[16] hinges on the possibility of finding *explicata* that satisfy these two requisites for each of our three *explicanda.*

Let us begin, then, by considering procedure.

It would seem that the quickest way to arrive at an *explicatum* of procedure is to start by analyzing in detail the stages into which a sample of scientific inquiry unambiguously taken as representative can be broken down. I shall follow this course, and resort to Galileo's explanation of sunspots in the Third Day of his *Dialogue Concerning the Two Chief World Systems.* The inquiry can be divided into four stages.

The first stage is that of "continual observations,"[17] that is, the systematic collection of data with the aim of finding out "the places from day to day when the sun was on the meridian."[18]

The second stage is that of "conjecture." In order to explain the trajectories observed, Galileo advanced the hypothesis that "such spots were of a material which was produced and dissolved within a brief time. As to their place, they were contiguous to the body of the sun, which revolves upon its own center in the space of nearly one month."[19]

There follows the third stage, which involves drawing the logical conclusions from the hypothesis advanced. If the hypothesis is true, then some "extraordinary changes would have to be seen by us in the apparent movements of the solar spots."[20] Galileo went on to list these changes with precision.

This brings us to the fourth stage, which consists in testing the "conjectured results" with new observational data. Galileo wrote: "It came about that, continuing to make very careful observations for many, many months, and noting with consummate accuracy the paths of the various spots at different times of the year, we found the results to accord exactly with the predictions."[21] Thanks to this correspondence, Galileo felt authorized to conclude that, even if his hypothesis was not absolutely certain, it was better than the Ptolemaic one.

The following sequence of operations could thus become a first approximation of an *explicatum* of the procedure followed by Galileo:

(1) $$O_i \ldots H_p \Rightarrow O_t \rightarrow C_t$$

where O_i indicates a set of initial data, H_p a plausible hypothesis which takes account of these data, O_t a set of testing data, C_t a final controlled claim; . . . is a link to be specified, $\Rightarrow$ is a deduction, and $\rightarrow$ is an induction.

Explicatum (1) is known as the hypothetico-deductive method. Since it is not difficult to demonstrate that this method is exemplified by several scientific inquiries in many different domains, and that many philosophers and scientists accept it despite their own varying epistemological commitments,[22] I believe I can be exonerated from providing further examples and safely consider it a good candidate.

However, it is by no means the only one. Consider the following *explicatum:*

(2) $$O \rightarrow C$$

where O is a set of observed data, C is a cognitive, explanatory claim, and $\rightarrow$ is an induction.

Although (2) might appear to be a segment of (1), there are authors who take it as a complete and legitimate method in itself, at least in certain cases. It has the same structure as the old *ars inveniendi,* for example, Bacon's inductive method (where $\rightarrow$ represents the inferences obtained through the inductive tables). Following Mill, we might call this a *direct (inductive) method.* Significant samples of scientific inquiry are covered by it. It is exemplified, for example, in the First Day of Galileo's *Dialogue* when, from certain observational data, he draws the conclusion that there is no essential difference between celestial and terrestrial regions. Clinical medicine is full of situations in which the application of certain inferential techniques to a set of symptoms allows the doctor to draw pathogenetic conclusions (so-called "differential diagnosis"). This is also true for other fields of inquiry. Furthermore, *explicatum* (2) seems to agree with several influential methodological views, such as Newton's "analytical method" (or "deduction from phenomena").

There are yet more *explicata,* such as the following:

(3) $$P_1 \ldots TT \Rightarrow EE \ldots P_2$$

where P_1 is an initial problem, TT a tentative theory, EE an attempt to eliminate error, and P_2 a final problem; $\Rightarrow$ is a logical deduction and $\rightarrow$ a link to be defined. This *explicatum* is Popper's well-known *method of conjectures and refutations.* Due to its general applicability, it is also easily exemplified.

Two problems crop up. The first is whether *explicata* (1), (2), and (3) are not redundant, that is, different or elliptical formulations of the same schema. Now it may well be that the difference between (1) and (3) is not methodological but epistemological, and related to Popper's well-known idiosyncrasy regarding induction. Nevertheless, (2) is certainly different from (1) and (3) on at least one essential point: while (1) and (3) are *consequentialist* methods requiring two different kinds of facts (initial facts on which the hypothesis is constructed, and final facts, or "novel facts," with which it is tested), (2) is a *generativist* method where the justification of the cognitive claim requires only one class of facts, as long as there is a logical link between the facts and the claim.

The second problem to be solved is which of these and other *explicata* better satisfies our two requisites.

Let us begin with the requisite of adequacy. It is hard to claim that there is *a single* answer to the problem, especially if one looks at the history of science without blinkers on. Significant scientific changes have always gone hand in hand with methodological innovation, and, since procedures are directed towards specific aims, a change in aim may well alter the procedure.[23] Galileo did not just change Aristotle's physics, he also altered his method (procedure). The same could be said for Darwin and many other innovators. The hypothetico-deductive method, for example, was advanced after theoretical entities were introduced into physics, and it was not fully accepted until this introduction brought about significant results.[24] Different epochs have different methods; likewise, different disciplines in the same period may have different methods. When physics was already mature enough to adopt the hypothetico-deductive method, other disciplines were still Baconian.

Let us be optimistic and suppose that a single, adequate *explicatum* can be found for all the main models of scientific practice. *Explicatum* (3), due to its general applicability, seems best to serve this purpose. The question now is: how precise is it?

By way of example, let us consider the following dialogue between a Guru and his pupil, taken from a Chinese acupuncture manual:

> *Guru*. In the beginning there was the Yin and the Yang, the foundations of all acupuncture. . . . Chinese medicine is based on the fact that everything in nature is formed by the Yin and the Yang. They are opposite but complementary forces, in the sense that they are perpetually in motion and they react with one another.
>
> *Françoise*. What are the Yin and the Yang really?
>
> *Guru*. The question should not be posed in these terms. They are not defined entities. Do not consider them as nouns, think of them as adjectives. In the positive-negative opposition, for example, Yang is positive and Yin is negative. . . . The whole structure of the body derives from the Yin-Yang opposition. Your back is Yang and your abdomen is Yin. The left half of your body is Yang and your right half Yin. Organs are Yin and your guts are Yang. Everywhere, then, there is a balance between Yin and Yang. In fact, one cannot triumph over the other. When there is too much Yang, you get a fever; when, in contrast, there is too much Yin, you get the shivers. Whenever there is imbalance, there is also illness. The aim of acupuncture is therefore to re-establish a balance using procedures which I will teach you. . . . We could think of Yang as being pure energy and Yin as being condensed energy—that is, matter.
>
> *Françoise*. Have we done with energy?
>
> *Guru*. Not yet. . . . I think it would be interesting to talk to you about perverse energy. Oe energy is a war-like energy that presides over the defense of your organism. Against what aggressors? Against perverse energies. When war breaks out, there is a fever, and the invalid fears the cold. When Oe energy triumphs, the invalid sweats and perverse energy is dissipated.
>
> *Françoise*. So perverse energy means illness?
>
> *Guru*. Not exactly. Illness is the consequence of aggression by perverse energies. There are five in all: wind, heat, damp, dryness, and cold. When wind penetrates the meridian muscles and tendons it provokes sneezes, coughs, and rhino-laryngitis. If the Oe defense is inadequate, wind penetrates the main meridians, causing fever, sweating, and headaches.[25]

I have no intention of discussing here whether Chinese acupuncture is scientific or not, though the above dialogue discourages even the most open mind. The most we can do is associate ourselves with Françoise's skepticism and observe that few people would be prepared to deny that the etiopathogenetic "explanations" provided by the Guru are very vague and based on extremely poor evidence. Nevertheless — and this is the point that interests us — it is not in terms of procedure that the Guru's explanations and evidence can be criticized. Indeed, if Françoise were interested in methodology and wanted to put her Guru into a tight spot, he could easily retort that like any good theoretician of experimental medicine in the West he introduces hypotheses and provides observational evidence to support them. Thus, the procedure used by the Guru to justify his cognitive claims follows all four steps of *explicatum* (3).

The same could be said for other kinds of inquiry. Freud always claimed, for example, that psychoanalysis — like any other natural science — follows the steps of (1) or (3) in both therapy and theory.[26] It would not be difficult to show, moreover, that this would also be true for many an inquiry in the fields of alchemy, natural magic, or astrology.

Let us forgo further proof, then, and come to a conclusion instead. Since the procedure we chose as the most adequate turns out to be so imprecise that it even manages to save inquiries which we consider pseudo-scientific, we may express this conclusion in the form of a paradox which I shall call *the paradox of scientific procedure.* It states: *given an adequate scientific procedure, it is possible to find inquiries considered pseudo-scientific which will satisfy that procedure.*

This paradox is alarming. It shows that method — in the sense of procedure — cannot successfully become a criterion for demarcation between science and pseudo-science, either because it is not sufficiently universal or because it is not sufficiently precise. The accusations against method put forward by supporters of the counter-methodological model have thus struck home, although a defender of the Cartesian project may still see no reason to relent. He could still hope to carry out the Cartesian project by taking into account not only the procedure but the techniques and rules as well. Let us see if he succeeds, starting with techniques.

1.2 Techniques

At first sight, relying on techniques in order to establish whether or not a given inquiry or discipline possesses a scientific status seems quite promising. To start with, it is a well-trodden path. Let us consider a few instances.

First example. One of the most frequent criticisms advanced by orthodox clinicians against so-called "alternative medicine" is contained in the following passage:

A basic methodological fault of homeopathy is the way in which observations are collected. Often the most elementary requisites such as completeness, neutrality, and objectivity—indispensable for the drafting of any scientific principle—are missing. Thus, for example, homeopaths hardly ever provide quantitative data when reporting their therapeutic and clinical experiences; nor do they try to make their observations repeatable.[27]

In this instance, it is clearly not the procedure that is being criticized but the techniques for collecting data. Although homeopathy may respect the standard procedure, the two clinicians reason, its techniques are not correct because they do not satisfy certain requirements.

Second example. It is now the turn of psychoanalysis to be reprimanded. One critic objects that analytical hypotheses are "never tested by psychoanalysts with the help of the standard testing techniques."[28] Another observes:

Suffice it to remember that clinical work is often very productive of theories and hypotheses, but weak on proof and verification; that in fact the clinical method by itself cannot produce such proof because investigations are carried out for the avowed purpose of aiding the patient, not of putting searching questions to nature. Even when a special experiment is carefully planned to test the adequacy of a given hypothesis there often arise almost insuperable difficulties in ruling out irrelevant factors, and in isolating the desired effect; in clinical work such isolation is all but impossible.[29]

A third example could be useful. The reference this time is to philosophical psychology. One critic points out:

Until a century ago, all attempts made by philosophers to construct psychological systems to explain how our minds work were mostly based on introspection, in the sense that they relied on "internal" factors such as feelings and thoughts. When psychology tried to establish itself as a separate scientific discipline, the first step to take seemed to be to adopt an experimental method according to the successfully tested paradigm of the natural sciences. Introspection, in fact, could never be a method, or procedure, for collecting experimental data which would comply with this paradigm.[30]

It is by now evident that, in these three cases, references to "method" and "procedure" have really more to do with techniques. The "methodological fault"—to quote the first passage—is always the same. Our critics hold that homeopathy, psychoanalysis, and philosophical psychology are not scientific because they use unreliable techniques.

Why not, then, appeal to technique for defining and demarcating science? Following this path, we might be able to find, if not a general methodological rule of demarcation, at least a disciplinary criterion for specific domains. One could attempt, for example, to define the "scientific virtue" of physics on the basis of its increasing reliance on mathematical techniques. Several authoritative interpretations about the birth of modern science, from Kant onwards, have stressed that the scientific nature of a discipline has often been linked with mathematics.[31]

It is a pity that the more attractive this path appears, the shorter the distance one can travel along it. Two considerations reveal that it allows us to take a few steps forward but comes to a dead end.

In the first place, there are two kinds of techniques: specific and domain-dependent techniques (e.g., the double-blind technique in medicine, the Rorschach technique in psychology); and general techniques that are wholly or partly indifferent to the domain in which they are employed—principally mathematical techniques. As far as the first kind is concerned, we cannot oblige a discipline to adopt one or more specific techniques when they are instruments invented by scientists during their inquiries and often change in the course of the inquiry. To link the scientific status of a discipline to a particular technique would be tantamount to arresting the progress of that discipline. As for the second kind, although by definition they guarantee objectivity and rigor, it is arbitrary to maintain that such disciplines as biology, economics, and geology became sciences only when they started relying on mathematics. The ways of objectivity and rigor are fewer than the ways of the Lord, which are reputed to be infinite; nevertheless, there are several to choose from and it would be methodological arrogance to prescribe one over another.

We must take note, then, of the fact that the category of admissible techniques is open and that there is no reason to limit it. This proves that if we look for an explication of scientific method in terms of techniques, and transform them into a demarcation criterion for specific disciplines, such an explication might perhaps be faithful (as a photograph of an individual would be at a given moment) but it could turn out to be useless (as the same photograph would be if it were used to identify the same individual at a later date). This is precisely what the prosecution has stated in the trial against science.

In the second place, it is already clear that the real problem concerning techniques is that of the *criteria* according to which they are used. Techniques are not good or bad in themselves. As instruments, they are only good or bad according to the aims and the ways in which they are employed. There are modern astrological texts that contain more graphs and formulae than a book of physics. Does that mean that astrology has become as scientific as physics? No, it just means that it is not the beard that makes the philosopher. If we want to label a discipline scientific, we must do more than examine the techniques used; we must examine *how* they are used.

I can clarify this point with a few examples.

Consider the case of psychoanalysis. The main objection to analytical techniques for collecting data does not regard the techniques themselves but the way the collected data are put to use. What some people object to is that clinical data do not give genuine support to psychoanalytical hypotheses because the data are incomplete, vague, or contaminated by interpretations. Objections would not be raised if Freud had given only a heuristic value to data obtained through analytical sessions and if he had then used independent tests such as epidemiological ones.[32]

The case of introspective psychology is no different. What is it that distinguishes it from scientific psychology? Appealing to Jean Piaget, one could claim that no epistemically significant difference has anything to do with the technique of introspection: "the only systematic difference . . . is a difference of method. The scientific psychologist, even when he introspects, is interested in verification."[33] This is true in other cases. Homeopathy does not distinguish itself from clinical medicine by "the way in which observations are collected," but by the way they are used. Likewise, philosophical sociology is not different from empirical sociology because it uses the *Verstehen* technique, but because it uses it wrongly, namely, evidentially, not just heuristically (just as philosophical psychology uses the technique of introspection evidentially).[34]

If we decide to generalize, we arrive at another paradox which I shall call *the paradox of scientific techniques.* It states: *a scientific discipline can legitimately adopt the same techniques used by pseudo-scientific disciplines.*

Thus, our attempt to find a satisfactory *explicatum* of "scientific technique" to be used for the Cartesian project has failed. At this point, if this project is ever to get off the ground, it will have to start from a different basis. As in a well-conducted trial, the next step can only be to call forth the last defense witness: rules.

1.3 Rules

We can illustrate what we normally expect of rules with the example of psychoanalysis. From Hans Eysenck's critical analysis,[35] one can select, among many others, the following objections.

1. Freud does not provide us with experimental evidence of any kind;

he relies on anecdotal evidence of the most unreliable variety. It is second-hand, selective and incomplete.

2. The original statements are couched in such vague, general, and complex ways that deductions cannot be made with any degree of definiteness.

3. Concepts like reaction formation are essentially ad hoc hypotheses which explain the case at hand because they have been put forward in order to explain it although they do not fit into any systematic framework.

After a series of objections of this kind, Eysenck sums up: "What is wrong with psychoanalysis? — [It] is simple: Psychoanalysis is unscientific. It is only by bringing to bear the traditional methods of scientific inference and experimentation that we can hope to reap all the benefit of its founder's genius."[36]

But matters are not so simple. Adolf Grünbaum, while criticizing psychoanalysis, vindicates it of many of the charges put forward by Eysenck, in particular that it is not falsifiable. But this is not the crux here. The point is that Eysenck, in the passages quoted, refers to "methods" in the sense of rules, and claims that a discipline can only be called scientific if it follows certain rules. Let us try and understand whether this is really true, and whether the Cartesian project can be saved if we follow this path.

A set of methodological rules constitutes what may be called a *scientific code*. Although these rules are often numerous — Lakatos's "sophisticated" version of Popper's methodology, for instance, contains a considerable number — they can be reduced to at least three fundamental rules. Their schemes are:

Acceptance rule (AR). A cognitive claim must satisfy such and such requisites in order to be recognized as part of the body of scientific knowledge.

Rejection rule (RR). A cognitive claim will be rejected for such and such reasons.

Preference rule (PR). A cognitive claim will be preferred to a rival claim if its satisfies such and such properties.[37]

Our first question is: how can we fill the "such and such" clauses? That is to say, what *explicata* can be provided for the basic rules of a scientific code?

The best way to find good candidates (as we have seen when we examined the matter of procedure) is to start with a representative sample of scientific inquiry. Once again, to save time, I shall limit myself to one example which I shall use as paradigmatic.

Galileo, who has already been called as a witness for procedure, will be called again to testify for rules. Three passages are particularly useful for extracting *explicata* of rules AR, RR, and PR.

The first passage comes from a letter to Gallanzone Gallanzoni, dated

July 16, 1611. In this letter, Galileo discusses the hypothesis put forward by Father Christopher Clavius and others to restore the hierarchy between the lunar and the sublunar worlds by telescopic observations. The moon is said to be a "a highly transparent enclosure, like crystal or diamond, completely imperceptible to human senses; this surrounding fills the cavities and caps the tallest lunar eminences, it girdles that first, visible body, producing a smooth, crystal-clear, spherical surface which does not block the sun's rays."[38] Galileo raises the following objection:

> Truly, the image is beautiful; the only thing missing is that it has not been, nor can it be, demonstrated. Who cannot see that this is pure and arbitrary fiction that states nothing, only suggests a mere lack of contradiction? If the chimeras of our brains could bring about actions in the course of nature, I would be allowed to say, with as much authority, that the earth is a perfectly smooth, spherical surface; meaning by earth not only this opaque body where the sun's rays terminate, but together with this the diaphanous atmosphere that fills all its valleys, and stretches as high as the tallest peaks of its mountains, surrounding it spherically.[39]

In the second passage—taken from the Second Day of his *Dialogue*—Galileo considers the plausibility of the Copernican hypothesis. Having listed seven "reasons which seem to favor the earth's motion,"[40] he puts these words into the mouth of his spokesman, Salviati:

> Up to this point, only the first and most general reasons have been mentioned which render it not entirely improbable that the daily rotation belongs to the earth rather than to the rest of the universe. For I understand very well that one single experiment or conclusive proof to the contrary would suffice to overthrow both these and a great many other probable arguments.[41]

In the third passage—from a private answer to Cardinal Bellarmine regarding the cardinal's letter to Father Foscarini about the relationship between the Copernican theory and Holy Scripture—Galileo compares Copernican and Ptolemaic systems:

> It is true that it is not the same to show that one can save appearances with the earth's motions and the sun's stability, and to demonstrate that these hypotheses are really true in nature. But it is equally true, or even more so, that one cannot account for such appearances with the other commonly accepted system. The latter is undoubtedly false, while it is clear that the former, which can account for them,

may be true. Nor can one, or should one seek any greater truth in a position than that it corresponds with all particular appearances.[42]

We can now suggest the following *explicata* for the three fundamental rules that lie behind these three passages.

AR.1 Only those hypotheses testable through observational data are to be accepted.

RR.1 Any hypothesis whose observational consequences are contradicted by empirical facts is to be rejected.

PR.1 If two hypotheses clash, the one that explains more facts is to be preferred.

As we have taken Galileo's inquiry as paradigmatic, we can consider AR.1, RR.1, and PR.1—at least for now—as typical *explicata* for the fundamental rules of the scientific code. What is to be examined, then, is whether these *explicata* satisfy our requisites or whether there are others that work better. Our hopes for defending scientific method, and the Cartesian project with it, all rest on this point.

Are these hopes well founded? The suspicion that they are not arises immediately. First, because a quick look at the prescriptive content of these *explicata* is enough to show that they are far from precise. Second, because it is doubtful whether their precision can be improved without affecting their adequacy. A new paradox is in the air.

1.4 The Paradox of Scientific Method

Let us begin with *explicatum* AR.1. It is clearly not very precise in its key points. What does "testable" mean? What exactly are "observational data"? Galileo again comes to the rescue. From his objection that Father Clavius's hypothesis was an "arbitrary fiction that states nothing, [that] only suggests a mere lack of contradiction," one could say that he thinks this hypothesis can neither be confirmed nor falsified. AR.1 could therefore be substituted with the *explicatum*,

AR.2 Only those hypotheses that are confirmable by observational data are to be accepted;

or with the *explicatum*,

AR.3 Only those hypotheses that are falsifiable by observational data are to be accepted.

These two *explicata* may be more precise than AR.1, but—apart from the fact that their prescriptive content is different—it is unlikely that they are highly precise. In AR.3, for instance, "falsifiable" could be taken to mean either "logically falsifiable" or "actually falsifiable," as Popper showed in his attempts to define his falsificationist criterion of demarcation. In the first interpretation, the rule is precise enough but patently inadequate because it is *too wide;* its liberality embraces nearly all kinds of inquiry, from astrology and psychoanalysis to physics. In the second interpretation, the rule is precise only if it is integrated with what Popper calls "anti-conventionalist counter-moves" such as a ban on using ad hoc hypotheses. It is highly unlikely that a detailed list of these counter-moves could be drawn up, and even less likely that such a list would be desirable. In any case, it would make the rule inadequate because it is *too narrow.*

Similar considerations could be made for RR.1. A more precise *explicatum* could be:

RR.2 Any hypothesis whose observational consequences are contradicted by consolidated observational data is to be rejected.

In the second passage quoted in the previous section, Galileo seems to have this *explicatum* in mind when he states that "one single experiment or conclusive proof to the contrary" would be enough to overthrow the Copernican theory. Yet other *explicata* are possible and defensible. In this regard, Galileo once again proves to be a treasure trove of information.

In a passage from his letter to Francesco Ingoli of 1624, he discusses the anti-Copernican objection (which Ingoli had taken from Tycho Brahe) according to which "the eccentricities of Mars and Venus are different from what Copernicus assumed, and likewise Venus's apogee is not stable as he believed."[43]

Galileo comments:

> Here it seems to me you want to imitate the man who wanted to tear his house to the foundations because the chimney made too much smoke, saying that it was uninhabitable and architecturally flawed; and he would have done it had not a friend of his pointed out that it was enough to fix the chimney without tearing down the rest. So I say to you, Mr. Ingoli: given that Copernicus went astray with regard to eccentricity and apogee, let us revise this, which has nothing to do with the foundations and the essential structure of the whole system. If other ancient astronomers had had your attitude, namely to tear down all that had been built every time one found a particular detail that did not correspond to a previous hypothesis, not only Ptolemy's great edifice would never have been built, but one would have remained always without a roof and in the dark about heavenly matters.[44]

The rejection rule we can reconstruct from this passage is definitely more "sophisticated" than RR.2. It is interesting to note that whereas Ingoli advocates a *literal* application of RR.2,[45] Galileo offers an *elastic* interpretation. The new *explicatum* could thus be formulated in the following terms:

RR.3 Any hypothesis whose observational consequences are contradicted by established observational data, unless they constitute a local or secondary anomaly, is to be rejected.

Notice that by adopting RR.3, Galileo implicitly admits that ad hoc hypotheses can be introduced in order to save a theory undermined by recalcitrant facts. This contrasts with one of the possible interpretations of AR.3 and therefore poses a serious problem of consistency within the scientific code. I shall return to this problem in Chapter 3. It suffices here to observe that, while *explicatum* RR.2 is fairly precise (though not always adequate for Galileo's own scientific practice), *explicatum* RR.3 is vague and leaves a significant margin of discretionality. When should a fact that contradicts a theory be considered a mere "secondary anomaly"? When should it be considered a good reason for rejecting a hypothesis? How many anomalies, or what kind of anomalies, are needed for a hypothesis to be rejected?

The preference rule raises exactly the same kind of problems. A slightly more precise version of *explicatum* PR.1 that better fits Galileo's practice described in the third passage in the previous section, could be:

PR.2 If two hypotheses are rivals, the one with greater empirical content is to be preferred.

The following would also fit the bill:

PR.3 If two hypotheses are rivals, the one with greater excess of empirical content is to be preferred.

This rule would cover, for example, Galileo's argument in favor of the Copernican system after it had been confirmed by new telescope observations ("O Nicholas Copernicus, what a pleasure it would have been for you to see this part of your system confirmed by so clear an experiment!").[46] But even in this case, the two *explicata*—themselves along different lines—require further specification. "Empirical content" or "excess of empirical content" are not, after all, precise notions.

One could object that more "sophisticated" *explicata* may exist which would eliminate, or at least reduce to an acceptable level, the vagueness that surrounds the fundamental rules of the scientific code. Let us concede that this is the case; we shall see later to what extent. It is enough to say here that these moves cannot go beyond a certain limit: if Galileo had kept to a highly sophisticated and precise code, his inquiry would have ended in a straitjacket. The fact that it went ahead was because Galileo preached one thing while practicing another. He preached RR.2, declaring that "one single experiment . . . to the contrary would suffice to overthrow" the Copernican theory; but as soon as RR.2 went against it he adopted the more tolerant *explicatum* RR.3.

Feyerabend considered this behavior typical of the opportunism of scientists. Indeed, it shows that scientists not only allow themselves an ample margin of discretion in the application of the rules of the scientific code, they also appropriate the right to suspend the validity of the rules. We shall see later what the effects of this margin and this right can be. For now we can say that they provide sufficient evidence for another paradox which I shall call *the paradox of scientific rules,* namely, *given any methodological rule, there are always scientific inquiries in which it is violated.*[47] This paradox casts further doubts on the idea of scientific method and the Cartesian project as a whole.

What should we conclude? We have tried to find three adequate and precise *explicata* for the three *explicanda* of scientific method (procedure, techniques, rules). Instead we ran into three paradoxes. The first lesson we have learned is that when the precision of the *explicata* is increased, their adequacy is reduced, and vice versa. The case of rules illustrates this point clearly: RR.2 does not save Galileo's attitude towards the Copernican theory, while RR.3, which does, is much less precise. The second lesson we have learned is that scientists work with vague rules and assert their right to do so. Although it would be over-hasty to conclude that any different factors (subjective, cultural, social, etc.) can be sifted from this vagueness, as supporters of the counter-methodological model would claim, we have sufficient grounds for admitting that scientific codes contain gaps. Summing up these results we obtain what I shall call *the paradox of scientific method,* namely, *science is characterized by scientific method, but a precise characterization of scientific method destroys science.*

This paradox expresses an intrinsic limitation of every scientific code. It is something like a *principle of methodological indeterminacy:* adequacy and precision are two properties of scientific method whose product cannot go below a certain limit. But if adequate and precise *explicata* for scientific method cannot be brought to court, has science lost its case?

Our results suggest that the first thesis of the Cartesian project is untenable. We have seen that when subjected to analysis and compared with significant examples of scientific practice, the idea that a universal and precise method exists for distinguishing science from nonscience is not very realistic and can be pernicious to scientific progress. Scientific practice shows that there is *more than one* procedure and *more than one* set of rules, each with differing levels of adequacy and precision. Moreover, we have seen that methodological rules, which carry the greatest weight in the Cartesian project, contain significant lacunae due to their vagueness. Since these lacunae can only be filled with decisions (not necessarily based on taste), our results also suggest that the second thesis of the Cartesian project is equally untenable. As far as the third thesis is concerned, we have not yet reached any conclusions, and we shall come back to the matter later.

Before raising the white flag and admitting that the idea of method is doomed to failure, it is tempting to consider another line of defense. Maybe our criticism hit the wrong target. In order to fulfil his project, Descartes sought universal rules and tried to justify them as "inborn principles" of the human mind. It is probable that the Cartesian intention was commendable but that its actualization was flawed. Maybe one should look not for universal rules, consisting of a few general yet poor commands, but for *local* rules. Rather than proceeding a priori and deriving the method from allegedly fixed properties of reason or the mind, one could procede a posteriori, taking method as any other intellectual instrument and measuring its performance in relation to its objectives, hence resorting to the history of science or to scientific practice. Who can say whether this path would not allow us to straighten out the destiny of method? I shall explore this possibility in the next chapter.

From Method to Rhetoric

In our day, philosophical criticism alone is honored. The art of "topics," far from being given first place in the curriculum, is utterly disregarded.

G. B. Vico, *On the Study Method of Our Time*

2.1 Method and the History of Science. Inductive Justification.

Since we are at a crossroads and about to change paths, it is useful to consider where the old road has taken us. Let us retrace our steps a little way.

Strictly speaking, the paradox of scientific method does not prove that science is without method; it only proves that science has a spectrum of possible methods at its disposal, each of which might be adequate for a given discipline with a given goal at a given time. On the other hand, the principle of methodological indeterminacy reveals that for each method adequacy and precision cannot go beyond certain *limits*. It does not prove that these limits are so narrow that they turn method into mere "verbal ornament." Judging from our results so far, it would be mere provocation to conclude that science does not follow a procedure, that it is not governed by rules, or that all rules are equally good.[1]

But if methodology is not to meet with total failure, the spectrum of possible methods must be reduced to a reasonable number and the limit of indeterminacy must be brought to an acceptable value. This is the main problem of meta-methodology that I shall be dealing with in this chapter: the problem of choice between rival methods.

If we abandon the idea that there is such a thing as an ahistorical entity such as "reason" or the "mind," that contains the natural, universal principles of inquiry, the suggestion made at the end of chapter 1 of leaving aside Descartes' a priori method and using the history of science and scientific practice in order to find the best method seems to be quite reasonable. After all, scientific methods are intellectual instruments for achieving certain cognitive ends, and therefore the actual history and the concrete practice of science can provide us with useful information about which method to adopt and which ends to strive towards. Didn't we ourselves call history to our aid when we used Galileo's scientific research as a paradigm for scientific inquiry in general?[2] Why not tread this path more systematically and examine both history and scientific practice in order to find a method?

This was Lakatos's starting-point. He maintained that "all methodologies function as historiographical (or meta-historical) theories (or research programs) and can be criticized by criticizing the rational historical reconstructions to which they lead";[3] thus "history may be seen as a 'test' of its rational reconstructions."[4] Laudan followed the same lines, particularly as far as the methodological rules that govern theory change are concerned, starting from the "conviction that any theory of scientific change needs to accommodate the impressive body of evidence assembled by historians from the career of science itself."[5]

Let us examine this view more closely, taking *m* as a certain set of methodological rules and HOS as the history of science. There are two main ways in which we can use HOS to justify *m* and, hence, two kinds of historical meta-methodologies. In this section I shall be dealing with the first kind, which I shall call *inductivist historical meta-methodology.*

Inductivist historical meta-methodology starts with HOS and takes it as the empirical basis for deriving *m*. Naturally, *m* must satisfy the two minimal requirements of explication: it must be both adequate (that is, it must save at least the best scientific practice) and precise (that is, it must allow concrete epistemic judgments—for example, as to whether a research program is progressive or regressive). To this end, inductivist historical meta-methodology must accept the following two assumptions: (a) in order to obtain precise methodological indications, it must assume that HOS is highly homogeneous as regards scientists' fundamental appraisals of their own results; (b) in order for these indications to become constraints or to have heuristic value, it must assume that the practice of science as witnessed by HOS is uniform as regards the fundamental criteria of scientific rationality. Lakatos accepted the first assumption when he wrote that "while there has been little agreement concerning a *universal* criterion for the scientific character of theories, there has been considerable agreement over the last two centuries concerning *single* achievements."[6] In a similar vein, Laudan claims that "there is a widely held set of normative judgements . . . clearer and more firmly rooted than any of our overt and explicit theories about rationality in the abstract."[7] As for the second assumption, Laudan is inclined to take it as part of his "naturalized methodology," but Lakatos rejected it and rigidly separated methodology from heuristics.[8]

Strictly speaking, neither of these two assumptions can be proved. The first is an act of faith of a different nature but with the same underlying idea as that on which the Cartesian aprioristic methodology is based: the homogeneity of the history of science, or — as far as Laudan is concerned — the uniformity of normative judgments corresponding to the fixity of the mind or reason. This raises the suspicion that historical meta-methodology is as circular as Cartesian methodology: the former finds in the history of science the very method it favors, the latter finds in the mind (or in actual practice) those rules it considers most desirable. Methodology, in this case, is to history what Narcissus was to the lake,[9] and the whole project turns out to be a sort of sophisticated Cartesianism.

Under Lakatos's construal, inductivist meta-methodology consists in comparing *m* with a special class of protocols taken from HOS that express "the 'basic' appraisals of the scientific elite,"[10] or the "verdicts of the best scientists." Laudan's idea is similar. Picking up on a hint from Lakatos, he distinguishes HOS_1, "the chronologically ordered class of beliefs of former scientists," from HOS_2, "the descriptive and explanatory statements which historians make about science,"[11] and he proposes comparing *m* with HOS_1. More precisely, Laudan suggests comparing *m* with a subset PI (pre-analytical intuitions) of HOS_1 that contains "cases of theory acceptance and theory rejection about which most scientifically educated persons have strong (and similar) normative intuitions."[12]

In order to discuss this view, let us first establish what is being compared with what by distinguishing the following kinds of statements:

n = *normative judgments.* These are shared judgments regarding specific cognitive claims. They are Laudan's "pre-analytical intuitions," for example, "It was rational to accept Einstein's special theory of relativity in 1911."

p = *preference judgments.* These are personal judgments, also regarding specific cognitive claims. They are Lakatos's "'basic judgments' of leading scientists," for example, as Einstein would have said in 1905, "It is rational to accept the special theory of relativity."

V = *value judgments.* These are objective and general epistemic evaluations that make no mention of either specific cognitive claims or temporal indices. They are Lakatos's "value judgments" and Laudan's methodological standards, for example, "It is rational to accept well-confirmed theories."

N = *norms.* These are imperatives, commands, or suggestions as to how to conduct an inquiry. They are the "ought-statements" that constitute Lakatos's heuristics,[13] for example, "Accept only well-confirmed theories."

As it is the task of historical meta-methodology to select a group of V-judgments that accommodate historical evidence, let us begin by comparing V with n, as Laudan suggests. This procedure faces the following problems.

Dependence of normative judgments on value judgments. The first problem has to do with *whether* V-judgments can be derived from n-judgments. Let us suppose that historians and scientists agree on the status of a given theory T at time t compared with time t'. Let us imagine, for example, that they support n_1 = "It was rational to accept Copernicus' theory in 1837 when Bessel first determined stellar parallax, rather than in 1610 when there were only Galileo's telescopic observations." On what grounds can n_1 be expressed? There are clearly two factors at work: a comparison of the different status of the Copernican theory in 1610 and in 1837, *plus* a value judgment about this difference, such as "It is rational to accept empirically well-confirmed theories." But this means that every n-judgment depends on a tacitly or explicitly formulated V-judgment. Deriving, or basing V on n, then, is tantamount to deriving or basing it on itself. This procedure is circular (and narcissistic). We end up by discovering in the history of science precisely what we had already deposited in it, though in disguise or long forgotten (in the sense that an appraisal of a particular historical case is not always preceded by an explicit formulation of the values upon which the appraisal has been based).

Indeterminacy of value judgments underlying normative judgments. If we admit that a V-judgment always lies behind an n-judgment, the second problem that arises is: *which* value judgment exactly? Let us consider the following value judgment V_1: "It is rational to accept well-confirmed theories." A historian who shares this view can use it to express, say, the normative judgment n_1, "It was rational to accept Copernicus' theory in 1837," and the normative judgment n_2, "It was rational to accept Darwin's theory in 1900." Another historian might have a different view. He might agree with V_1 and accept that it covers n_1 but deny that it covers n_2. This is possible because, as we have said, a normative judgment on a theory depends not only on a certain value judgment but also on an analysis of the status of that theory at the time it was considered. But if one historian, in the name of V_1, expresses n_1 and n_2, and another historian expresses n_1 and not n_2, this means that V_1 is indeterminate or vague. The point here is not only that there is a value judgment behind every normative judgment, but that the very same value judgment can lurk behind different and even incompatible normative judgments.

Underdetermination of value judgments by normative judgments. A third problem still remains: *how many* value judgments can be derived from normative intuitions? Let us consider the following example. Scientists and historians of science all agree that in 1911, at the time of the first Solvay Conference in Brussels, nearly all the most famous physicists accepted the special theory of relativity, or at least its mathematical formalism. They generally also agree that it was rational for these physicists to accept the theory at that time. What they often disagree on is *why.* Some think it was rational to accept the theory because of its great heuristic power; others are of the opinion that it was rational to accept it because the theory was well-confirmed empirically; still others point to different reasons.[14] Such disagreement is by no means cause for scandal; it is, nevertheless, a source of serious problems for those who want to compare *V*-judgments with *n*-judgments and base the former on the latter. It proves that even when there is agreement on a certain set of *n*-judgments, this is not enough to determine univocally a single *V*-judgment, because the former underdetermines the latter. It would be absolutely mistaken to claim that *any n*-judgment is compatible with *any V*-judgment. It is enough for one *n*-judgment to be compatible with more than one *V*-judgment to show that the former are not a stable enough foundation for the latter.

The same problems arise when *V* is compared with *p,* as Lakatos suggested. If anything, the situation here is even less favorable to inductivist historical meta-methodology.

In the first place, a *p*-appraisal such as: "It is rational to accept the special theory of relativity" can be used only to support a *V*-judgment such as "It is rational to accept theories with great heuristic power" if one believes that what Einstein did was really rational. But this takes for granted the very *V*-judgment in question, that is, that it is rational to accept theories with great heuristic power. We are locked in a circle: a *p*-appraisal confirms a *V*-judgment just because it depends on the *V*-judgment.

Second, for every *p*-appraisal there is nearly always an opposite *p'*-appraisal. The history of science contains both p_1, "[Einstein, 1905:] It is rational to accept the special theory of relativity," and p_2, "[Lorentz, 1905:] It is not rational to accept the special theory of relativity." On what basis should we decide that p_1 is a better basic judgment than p_2? Because Lorentz was in a minority? There has to be a criterion for selecting a class of *p* appraisals. If there is, we are again in a circle; if there is not, it must be admitted that the class of the verdicts of the scientific elite is so heterogeneous that the only value judgment that can be derived from, or better accommodate, those verdicts is necessarily ***vague and imprecise.***

Third, and finally, a class of *p*-appraisals does not support only one *V*-judgment, but many *V*-judgments, since the reasons underlying scientists' personal preferences can be, and usually are, numerous, conflictual, and not always recommendable. As Feyerabend has written, quite rightly, "basic value judgments are only rarely made for good reasons."[15] Einstein thought that it was rational to accept the special theory of relativity in 1905 because of its symmetry, elegance, and simplicity; Tolman and Lewis accepted it in 1908 on empirical grounds; Wien accepted it in 1909 for its logical consistency, and so on.[16] What *V*-judgment can be derived from, or adapted to, these "basic appraisals of the scientific elite"?

Having got thus far, we should briefly sum up the situation. Attempts made by inductivist historical meta-methodology to restrict the range of methods and diminish the product of their adequacy and precision meet with serious difficulties. Two in particular. Due to the underdetermination of value judgments by pre-analytical intuitions and by scientists' basic judgments, we are left with ***many*** methods. The paradox of scientific method raises its head again since the choice of just one of these methods may turn out to be inadequate for certain examples of scientific inquiry. Furthermore, because of the heterogeneous range of pre-analytical intuitions and basic judgments covered by the same value judgment, these methods will still be ***vague.*** There is, then, little reason for believing that methodological indeterminacy has been effectively reduced.

Inductivist historical meta-methodology faces other serious obstacles. The first has to do with the normative aspect of methodology. Let us assume that a well-selected sample of basic judgments or normative intuitions definitively determines a few precise *V*-judgments, such as Lakatos's value judgments of his methodology of research programs. This means that scientists in the past explicitly or implicitly adhered to these judgments. But why should it be rational to heed them today (or in the future) under different circumstances? Because it is a good thing to be conservative, perhaps?

As far as moral value judgments are concerned, this obstacle dates back to Socrates: is a saint a saint because he pleases the gods or does he please the gods because he is a saint? By analogy, the same question arises for scientific value judgments: is it rational to accept well-confirmed theories because Galileo, Newton, and Einstein did, or did Galileo, Newton, and Einstein accept them because it was the rational thing to do? In the first case, ***ought*** is inferred from ***is,*** and one ends up in the naturalistic fallacy; in the second case, ***ought*** is used properly to judge what ***is*** (or ***has been***), but the history of science is no help in making these judgments—except as a consolation. As we have already said, Lakatos proposed separating methodology (*V*-judgments) from heuristics (*N*-imperatives), but this solution is unacceptable because, if methodology offers criteria for distinguishing good science from pseudo-science, these criteria are themselves recommendations for those who intend to practice good science, or at least avoid pseudo-science.[17]

The second obstacle has to do with methodology considered independently of heuristics. The more basic judgments that are covered by a value judgment, the vaguer the value judgment gets. Even the most accurate methodological rules have an "open texture," because they can be used in a range of possible applications which is not and cannot be completely defined.[18] These rules contain lacunae, and since these lacunae can be filled only with case-by-case decisions (that is, every time a rule has to be applied to a concrete situation), an important premise of the Cartesian project (originally shared by Lakatos, too)[19]—that a rigorous application of methodological rules allows one to come to univocal conclusions—turns out to be no longer justified.

Lakatos took these obstacles into account: he ended by admitting that the idea that "there must be the constitutional authority of an *immutable statute law* . . . to distinguish between good and bad science"[20] is untenable, and recognized that it was "*hubris* to try to impose some *a priori* philosophy of science on the most advanced sciences."[21] This *hubris,* however, does not refer only to the original Cartesian project, or to Popper's project (the "Euclidean methodologies" Lakatos spoke of), it also refers to Lakatos's own sophisticated Cartesian project, which has the same goal as the others ("universal definitions of science," "sharp criteria," and so on). This reveals that the historical path on which Lakatos and Laudan embarked in order to justify their own methodologies may be longer than Descartes' but it does not necessarily take them further.[22]

One final consideration, however, is important. History-oriented philosophy of science has genuine merits. It was an important event when the formal approach of logical positivism was left by the wayside and when both the history and practice of science were taken into consideration. Philosophers came to discover that between heaven and earth (for which various terms were coined such as Popper's World 2 and World 3) there are many things whose richness escapes logical models. The fact that some—Lakatos for one—tried to force history into "reconstructions" as straitjacketed as the old models, and the fact that, having dealt with too many Hegelian "explosives,"[23] they cultivated the dream of reducing the reality of history to the rationality of its reconstructions does not detract from the advantages of the new approach. Problems which were previously neglected or unknown, such as theory change, intertheory relations, meaning changes, incommensurability, and rationality, have come to dominate the scene in philosophy of science. As far as methodology in its strictest sense is concerned, however, appeals to the history of science have yielded few significant insights.

Must we then conclude that the history of science is irrelevant to methodology? No. We should say, rather, that it is not essential. The history of science provides us with *examples* of rationality; it would be a mistake, however, to transform these examples into *foundations, guarantees,* or *justifications* of method. We cannot say that a method *m* is good because it saves HOS outright or in part; we should state that if *m* is good then that part of HOS that is saved by it is also good. We must always be ready to correct the judgments of the gods: first, because gods make mistakes and, second, because gods quarrel. But it would be impossible to correct the gods' mistakes and to put an end to their quarrels if we mistook their desires and caprices for dogmas.

2.2 Method and the History of Science Hypothetico-deductive Test

Once again it seems we are doomed to conclude that the methodological project is a failure. But once again, before raising the white flag, we should ask ourselves if we have not made a mistake somewhere along the road. Perhaps the error is not in the path we have followed (the history of science) but in the goals we were trying to achieve (a universal and precise methodology). After all, inductivist historical meta-methodology is still Cartesianism, although of a sophisticated kind. It might well be that Cartesianism is overambitious. As we have seen, it sets out to prove that science has only one goal and that there is only one method to achieve that goal. If we are to look at real science without doing violence to it, why can't we admit that this assumption is untenable? Why can't we openly recognize that history shows us several ends and means at work in science? Why can't we, then, change the premise and look for something less ambitious: rather than *the* universal method, at least *one* method which, given *one* end, is better than others?

The old thesis worked on the following assumption:

> Underlying historical changes of theory, there is, moreover, a constancy of logic and method, which unifies each scientific age with that which preceded it and that which is yet to follow. Such constancy comprises not merely the canons of formal deduction, but also those

criteria by which hypotheses are confronted with the test of experience and subjected to comparative evaluation.[24]

The new thesis states:

> The absence of a single overarching methodology for science only complicates the task of the methodologist, it does not undermine it. . . . The task of the methodologist of science, as I see it, is to formulate methodological rules for the realization of the various plausible ends of scientific inquiry which one finds in place in the scientific community. Once formulated, those methodological maxims — conditional on specific ends — have to be evaluated.[25]

I shall call this view neo-Cartesianism. It is Cartesianism because it still links science with method, but it is neither standard Cartesianism, because it denies a single universal method, nor sophisticated Cartesianism, because it looks not for an inductivist, but for a hypothetico-deductivist justification of methodology; or, to use Laudan's expression, for a "normative naturalism."[26] This historical meta-methodology has two versions according to whether past historical cases or current scientific practice are taken as evidence: I call the former *backward looking*, and the latter *forward looking*.

Considering method as a means to achieve certain ends means taking methodological rules as hypothetical imperatives: if your goal is g, then use m. Considering history or scientific practice as evidence means submitting a methodological hypothesis m to the kind of tests that are used for empirical hypotheses: if HOS proves that g can be achieved more easily using, say, m_1 than using m_2, then m_1 is more reliable than m_2. This was originally Popper's suggestion,[27] but only Laudan developed it systematically and transformed it into a philosophical research program. If the program works, the Cartesian project will certainly have to be revised but not rejected altogether. The first thesis will have to be reformulated since there is no longer a universal method; the second thesis, however, will stay as it is since each method will allow the goal it is devised for. Consequently, even the third thesis will be left intact, because the rationality of science will continue to be guaranteed by method. Let us try and examine in detail, then, whether the new program works, and how.

Taken as hypothetical imperatives, methodological rules (N-norms in our terminology) follow on logically from premises containing value judgments (V-judgments) and from empirical statements r describing means-end regularities. Popper's rule of submitting hypotheses to serious attempts at falsification, for example, would be taken as a suitable conclusion to an argument like the following:

V: It is rational to accept theories that are approximately true.
r: (HOS shows that) to submit hypotheses to serious attempts at falsification is a means for approximating truth.
N: Therefore, if you want to be rational, submit your hypotheses to serious attempts at falsification.

Here the first premise indicates the desired end, the second an empirically tested regularity between a certain means and that end, and the conclusion expresses the rule. Laudan's view is that if we construe methodological rules in this way, as rules or maxims "resting on claims about the empirical world," then a test of methodological hypotheses is as stringent as that of empirical hypotheses. "Thus," he concludes, "we have no need of a special meta-methodology of science; rather we can choose between rival methodologies in precisely the same way we choose between rival empirical theories of other sorts."[28] Let us see how this can be done.

Suppose that our goal is g_1 and that our methodological hypotheses for achieving it are m_1, m_2, and m_3. A sample of history of science might offer us regularities such as the following:

(a) m_1—— 80% g_1 and 20% $\neg g_1$
(b) m_2—— 60% g_1 and 40% $\neg g_1$
(c) m_3—— 40% g_1 and 60% $\neg g_1$

Looking at these regularities, can we conclude that m_1 is better than either m_2 or m_3? A few serious difficulties arise. The first is the composition of the sample. As for the test of empirical hypotheses, the sample is represented by experiments that are dictated by the logical consequences of such hypotheses. We say that if a certain hypothesis h is true, it must pass certain evidence e_1, e_2, e_3, etc. But as regards the test of methodological hypotheses the situation is different. Here we cannot claim with the same assurance that if a certain method m_1 is reliable, then it must save certain historical cases c_1, c_2, c_3, etc. since we do not know precisely *which* cases are relevant. This is especially true when the desired end of methodological hypothesis is far-reaching and vague. Suppose, for example, that the end of m_1 is g_1 = obtaining true theories. What sample of history of science would be relevant for testing m_1? Cases in which scientists expressly professed g_1? Cases in which scientists actually pursued g_1 without necessarily professing it? Both? The whole history of science, because scientists are supposed always to pursue the goal of obtaining true theories, whether they admit it or not?

This difficulty can be further clarified with an example from Laudan. Take the following methodological hypothesis:

M: "If one is seeking reliable theories, then one should avoid *ad hoc* modifications of the theories under consideration."[29]

Now take the following empirical hypothesis which is structurally analogous to *M:*

H: If one wants to aid recovery from such and such an illness, then one should use such and such a therapy.

One of the most effective ways of testing *H* is by Mill's method of difference. Say r = recovery from the illness and t = the therapy. The test is made by observing (or artificially reproducing) a case where r is present and another where it is absent and then applying the eliminative principle according to which if the case where a phenomenon is present differs from the one where it is absent in only one circumstance, then that circumstance is the cause (necessary condition) of the phenomenon. Let us suppose that a comparison between the two cases gives the following result:

$$t_1, t_2, t_3 \text{ —— } r$$
$$t_2, t_3 \text{ —— } \neg r$$

The same method transferred to *M* would give this result:

$$m_1, m_2, m_3 \text{ —— } g$$
$$m_2, m_3 \text{ —— } \neg g$$

On the grounds of the first result we conclude that t_1 is the cause of r. On the grounds of the second result, however, can we likewise conclude that m_1 is the only (or the best) means for achieving g? Certainly, ***provided*** that the second result can actually be obtained, that is, if cases where g is present and the case where g is absent can actually be observed, as cases where r is present or absent can be observed (or reproduced). But this result cannot be obtained, at least not with the same degree of certainty. We simply do not know which theories are reliable and which are not. All we know is that certain theories have passed the tests to which they have been submitted while others have failed; but this is not enough to provide us with methodological rigor. The fact that a theory handled with m_1 passed the tests while others handled with m_2, m_3, etc. did not, might have nothing to do with methods, depending rather on other factors such as the instruments used, the ability to design experiments, the opportunity of carrying them out, and so on. Thus, when Laudan states, with regard to *M,* "assume, for the sake of argument, that we have reasonably clear conceptions of the meanings of relevant terms in this rule; indeed, without them, the rule could never be tested by anyone's meta-methodology,"[30] he assumes something that cannot be conceded. Even if a certain meaning of "ad hoc modification" were to be agreed on, the meaning of "reliable theory" would still be open to discussion, as would the composition of the sample to be used in testing the corresponding methodological hypothesis.

Let us grant, however, that an ideal situation exists where it is known which methods are used for achieving which ends. There would still be another difficulty to deal with, concerning the time interval within which the sample must be examined. Let g = obtaining reliable theories, m_1 = never use ad hoc hypotheses, and m_2 = use ad hoc hypotheses provided they are potentially, although not presently, testable. Now, in the short run, m_1 may lead to g more frequently than m_2 (e.g., because m_1 forces one to accelerate the rate of theory change), but in the long run m_2 may turn out to be more efficient (e.g., because saving a program with ad hoc hypotheses could stimulate one to invent new experiments). Is there a rule for establishing how long a time interval should be?

It might be objected that this difficulty is the same as the one that affects the test of empirical hypotheses. If a hypothesis h does not pass the test of evidence e, there is no rule for establishing after how many falsifications h should be rejected; likewise, if a method m does not achieve a goal g, there is no meta-rule for establishing after how many unsuccessful attempts m should be abandoned. Similarly, just as, if h has passed some evidence e, there is no rule for establishing after how many confirmations h can be considered proved; likewise, if m has achieved g, there is no rule for establishing after how many successful attempts m can be considered reliable. In other words, just as an empirical hypothesis is underdetermined by its confirmations and never definitively eliminated by its falsifications, a methodological hypothesis (rule) is always underdetermined by its successes and never eliminated by its failures. And there seems to be no reason why a greater degree of certainty should be required of methodological hypotheses than of empirical hypotheses.

In practice, things are different. If an empirical hypothesis h_1 is logically underdetermined by evidence e, one could try to choose between h_1 and its equally underdetermined rivals h_2, h_3, etc. by using some methodological rule or other (e.g., a rule that refers to values—say simplicity).[31] But if a method m_1 is underdetermined by the historical cases used to test it, then historical meta-methodology has no other instruments at its disposal for discriminating between m_1 and its rivals m_2, m_3, etc. Thus, if one sets oneself a goal g_1 and finds that the relevant samples of history of science in which g_1 was achieved cannot discriminate between, say, Lakatos's methodology of theory change and Laudan's, then one must surely conclude that the history of science provides an inadequate instrument for reducing the range of possible methods.

A further difficulty compounds this. It concerns neither the composition of the sample of history of science nor the time limits for its use. It concerns, rather, the determination of the degree of warrant a given sample of the history of science can provide for a given methodological rule. Hypothetico-deductive historical meta-methodology, in this situation, seems to go round in a vicious circle, because in order to test methodological rules it is forced to assume the validity of the very rules of the test. Laudan, who is well aware of this difficulty, claims it could be avoided "provided that we can find some warranting or evidencing principle which all the disputing theories of methodology share in common."[32] This principle can only be an inductive principle, which Laudan formulates in the following terms:

> (R1) If actions of a particular sort, *m*, have consistently promoted certain cognitive ends, *e*, in the past, and rival actions, *n*, have failed to do so, then assume that future actions following the rule "if your aim is *e*, you ought to do *m*" are more likely to promote those ends than actions based on the rule "if your aim is *e*, you ought to do *n*."[33]

According to this principle, the problem of determining the value v of the degree of warrant w of a method m based on a sample s of history of science—in symbols: $w(m,s) = v$—becomes the same as the problem of determining the value r of the degree of confirmation c of a hypothesis h on the basis of evidence e—in symbols: $c(h,e) = r$. In both cases, it is a matter of inductive logic, of a first level in the latter case, and of a meta-level or second level in the former.

Let us deal initially with the first level, considering the case of *singular predictive inference,* i.e., that inference with which, having observed that a number of individuals enjoy a certain property, one concludes that the next individual will most likely enjoy the same property. How is r determined?

Carnap thought it reasonable to assume that r lay in the interval m/n, w/k, that is, in the interval between the observed empirical frequency of the property in question and its logical frequency. In Carnap's view, the exact point at which r can be placed is given by the arithmetic mean of these two frequencies, weighted with two measures, one equal to n, the other equal to a parameter lying between 0 and ∞ according to the formula:

$$c(h,e) = r = \frac{\dfrac{m \cdot n}{n} + \dfrac{w \cdot \lambda}{k}}{n + \lambda}$$

Let us now proceed to the second level. In this case, the singular predictive inference is the one with which, having observed that a certain method *m* has led to a goal *g* with a certain frequency, we conclude that the next use of *m* will most likely lead to *g*. If we apply Carnap's reasoning to this, we obtain the formula:

$$w(m,s) = v = \frac{\frac{p \cdot q}{q} + \frac{x \cdot \mu}{y}}{q + \mu}$$

where p/q is the empirical frequency of *m*'s successes; x/y is the logical frequency, which is equal to 1/2 if only two methods, that is, m and $\neg m$, are considered, and equal to $1/y$ if the y methods of the family to which m belongs are considered; and μ is the weight of the logical frequency.

It is clear that the values of r and of v depend on the values of λ and μ. First let us deal with λ. This parameter expresses the degree of uniformity we intend to attribute to nature. If we say $\lambda = 0$, our confidence in uniformity is maximum and we infer that at the next occasion the probability of the property under consideration coming up is equal to its observed empirical frequency. If, on the other hand, we take $\lambda = \infty$, our confidence in uniformity is minimal and we infer that the probability of the property coming up is equal to its logical frequency. If, finally, we take $0 < \lambda < \infty$, our confidence in uniformity is at an intermediate level and we infer that the probability of the property coming up at the next occasion will be closer to the empirical or the logical frequency according to the value chosen for λ. The same reasoning can be repeated for the parameter μ. For $\mu = 0$, we can conclude that *m* will lead to *g* with the same probability as the frequency of its past successes; for $\mu = \infty$, the probability is equal to its logical frequency; for $0 < \mu < \infty$, the probability lies somewhere in the middle.

The vital question now is: in what way can we justify the choice of a value for λ and μ? As regards λ there is an excellent argument for excluding the value ∞. This is a transcendental argument, and Carnap resorted to it when he imposed on function *c* the condition of learning from experience. As for μ, Laudan seems to be referring to the same kind of argument when he states that "if (R1) is not sound, no general rule is,"[34] because if one does not accept (R1) one cannot learn from past means-ends regularities.[35]

That this argument proves a great deal—in particular that λ and μ must be different from ∞ if we want to learn from experience—is without doubt. Unfortunately, it does not prove enough. We have to assume *a priori* a certain uniformity, and yet we have no way *a priori* of assuming a *particular degree* of uniformity in the *continuum* of possible values. Thus, having used a transcendental argument for excluding that $\lambda = \infty$ and $\mu = \infty$, we are free to choose any one of their infinite values. This means that even a high frequency of cases in which a given goal *g* is achieved through a given method *m* still leaves ample margins for freely choosing a different method on the next occasion.

Since what has been said is also true for the *forward looking* version of the hypothetico-deductivist historical meta-methodology, we can draw this discussion to a close.[36] None of the above considerations proves—nor really intends to prove—that the history of science is irrelevant to methodology. Far from it. Especially when methodological rules are construed as hypothetical imperatives, historical and factual surveys often carry weight in the choice of rules (just as empirical evidence carries weight in the choice of theories). These considerations do, however, prove—and intend to prove—that this weight is not compelling, and even when it is it depends on decisions that cannot be imposed by any meta-rule. The same factors that through the choice of a particular value of λ come into first-level inductions also come into second-level inductions through the choice of a particular value for μ.

Of course, one learns from failures and successes, and we must not overlook past experience, but there are many ways of learning from experience and history (if one is interested in this kind of learning). Usually, if we want to achieve a goal, it is rational to use those methods which have proved efficient in the past; but occasionally it might be just as rational to change methods. This often happens in life, in business, even in love, with surprisingly new and better results. Likewise for rules. If Galileo had used only the most efficient rules of his day, modern science would never have been born. If Darwin had followed Bacon's standards, considered the most efficient at the time, we would still believe in the biblical version of Creation. If Einstein had not been an opportunist, betraying the canons of empiricist methodology, we would not have relativity theory, and, similarly, quantum physics would never have come about if a generation of physicists had not committed patricide against Newton and his rules. Granted, these are not definitive results, but few people would claim they are not important results and nobody would claim they are not rational. And yet, to admit they are is tantamount to admitting that one can be rational even while violating the most rational methods.

In the previous chapter, we sought *the* definitive scientific method that would be both adequate and precise. The paradox of scientific method and the principle of methodological indeterminacy have disclosed serious problems. In this chapter, we started out by taking a different path to reach the same place, but we did not meet fewer obstacles. So we decided to be a little less ambitious: instead of searching for *the* definitive method, we undertook to find *one* method for achieving *one* end. We would have been happy to give up the first thesis of the Cartesian project to save the second;

that is, we would have been pleased to have many different adequate and precise methods able to secure a given scientific goal rather than a universal method that is vague and inadequate. But we have not succeeded. If anything, we are in an even worse position because instead of dealing with a variety of methods we have to deal with a continuum.

So what lesson have we learned? That science has no method and that epistemic judgments and decisions cannot be rationally justified? *Yes,* if we agree with the third thesis of the Cartesian project, which states that without clear and precise rules science falls prey to "mob psychology." *No,* if we courageously throw out the third thesis along with the first and the second, and decide to replace the old idea of method with constraints of another kind, thereby freeing scientific rationality from slavery to methodological rules. But what other constraints?

2.3 From Method to Rhetoric. Back to Aristotle.

Let us first try to understand in more depth the lesson we have learned from our inquiry.

Just because our attempt to reduce the variety of possible methods failed, this does not mean that there are no constraints in science. Indeed there are: like any other form of experience organized according to goals and values, science has norms, habits, techniques, and practices with which it achieves its goals and values. Our failure means rather that any scientific code that is adequate and precise in a given situation cannot be adequate and precise in all situations. If one looks for rules, then, what one draws from the history of science or actual scientific practice are hypothetical imperatives such as:

> If you want to know a great deal about nature:
> (a) accept only well-tested theories;
> (b) reject systematically falsified theories;
> (c) choose theories that are richer than available ones.

Such rules, to use Kantian terminology, are *imperatives of prudence,*[37] that is, pragmatic and rather obvious recommendations. They are like the rules a prudent driver would wish to follow, for example:

> If you want to get home safe and sound:
> (a) drive at a moderate speed;
> (b) do not drink before driving;
> (c) stop if you feel sleepy.

Although these imperatives are constraints, they are far from precise. It is up to the driver, using his good sense, experience, and knowledge of the vehicle, to decide what speed is moderate, how much alcohol he can manage, and how tired he can safely get. Similarly, it is up to the judgment of the practicing scientist, guided by his education, to establish whether a theory is well-tested, a falsification systematic, a theory more promising than another, etc. When a methodologist tries to constrain this judgment by a detailed code of rules, the whole enterprise fails.

Do we have to resign ourselves to this? As we have seen, some, like Lakatos, who once believed in methodology, ultimately came to recognize that it was useless, gratuitous *hubris* even, to carry on trying. Taking this failure for granted, others supported a counter-methodological view and ended up by relegating the value of science to the domains of psychology and sociology. For them, once method is taken away, what remain are "subjective wishes" or "routine conversation," or "social conventions."

Too little, evidently. We must seek another path. The results of our analysis point to one in particular. Only by rejecting the third thesis of the Cartesian project can we be cured of that Cartesian syndrome that allows no alternative to method but irrationality. The ground to be covered is mostly new. We shouldn't attempt to eliminate subjective wishes and social conventions from science; rather we should try and incorporate them into science without sacrificing its undeniable nature of rigorous and objective knowledge. My claim is that this is possible provided we transfer science *from the kingdom of demonstration to the domain of argumentation,* and conceive its constraints not as universal methodological rules but as historical dialectical factors on which concrete interlocutors in concrete discussions rely.

This is the path we referred to in the Introduction as marked out by the odd (rabbit) pages of Kuhn; when he claims, for example, that the answer to how a scientific conversion takes place lies in "techniques of persuasion, and arguments and counterarguments, in a situation in which there can be no proof,"[38] or when he holds that "to discover how scientific revolutions are affected, we shall therefore have to examine not only the impact of nature and of logic, but also the techniques of persuasive argumentation effective within the quite special groups that constitute the community of scientists."[39] To a certain extent, this is also the path Rorty follows when he claims that "there are no constraints on inquiry save conversational ones—no wholesale constraints derived from the nature of the objects, or of the mind, or of language, but only those real constraints provided by the remarks of our fellow-inquirers."[40] Kuhn, however, has not gone beyond these important hints, and as for Rorty, he has never precisely defined the difference between, say, routine scientific and routine political conversation. On the contrary, by comparing the Galileo-Bellarmine case to that of Mirabeau-Louis XVI, he has claimed that "there is no epistemological difference between truth about what ought to be and truth about what is, nor any metaphysical difference between facts and values, nor any methodological difference between morality and science."[41]

This answer is plainly unsatisfactory. To say that persuasion is also present in science is not to deny any difference between science, politics, and ethics. We do not need experiments to be convinced that Churchill is better than Hitler; nor do we need calculations to be convinced that Mondrian is a better painter than a whitewasher. In both cases, what we have to look for is not the approximate genre but the specific difference, i.e. the typical forms of scientific persuasion. Without these, to say that science is "argumentation" or "conversation" (or to say that theory-change is a matter of "conversion") is as illuminating as stating that pens and computers are "objects for writing" or that men and amoebas are living organisms. In the blackest of nights, all cows are equally black, but this may have little to do with the color of cows and a great deal to do with the circumstances in which they are observed.

Before stepping onto our new path, I must clarify what I mean by "argumentation" and why I think it is useful to move science into the domain of argumentation. In the sense I give it, "argumentation" or "argumentative reasoning" is the same as "rhetorical argument." Although I shall often refer to Aristotle's authoritative view on the subject, I shall diverge from him on one important point.

According to Aristotle, a dialectical argument is a formally valid argument. The difference between a dialectical syllogism and a scientific syllogism is epistemic: the former starts off with authoritative premises (*éndoxa*), that is, theses that are "accepted by everyone or by the majority or by the wise—i.e. by all, or by the majority, or by the most notable and reputable of them";[42] the latter starts off with basic, true premises. On the other hand, Aristotle claims that rhetoric is a "counterpart,"[43] an "offshoot," or "branch," "similar" to dialectics.[44] It too starts off with *éndoxa* and uses the same or similar argumentative forms (enthymeme is a kind of syllogism, *apódeixis tis, syllogismós tis,*[45] example is a kind of "rhetorical induction," *epagogé rhetoriké*),[46] and it too belongs to the same class of theoretical arguments because "it is the function of one and the same art to discern the real and apparent means of persuasion, just as it is the function of dialectics to discern the real and apparent deduction."[47] According to Aristotle, the difference between dialectic and rhetorical arguments is practical. Dialectics is the art of attacking and refuting, while rhetoric is the art of persuading. A dialectician has before him an interlocutor who rebuts; a rhetorician, a silent audience (an assembly, or a jury, for example). A rhetorician has to be aware of the psychology of his audience, admittedly using the same arguments as the dialectician but in a more opportune manner, introducing, for example, his *éndoxa* not from the start but at the right moment, when he is sure they will produce the desired effect of persuasion. This is why rhetoric, unlike dialectics, makes use of such extralogical ingredients as the character of the speaker (*ethos*) and the passions of the audience (*pathos*).

This is the point where I depart from Aristotle. As in science, one cannot persuade an audience without attacking and refuting one's interlocutors and without defending one's own position, I shall incorporate Aristotle's dialectics into rhetoric and I shall call "rhetorical" those arguments that, in Perelman's words, aim "to induce or to decrease the mind's adherence to the theses presented for its assent."[48] But this is only a temporary characterization. Suffice it to stress here that, being a sort of *reasoning*, rhetorical arguments in my sense differ from nondiscursive techniques of persuasion (*atechnai pisteis*, nontechnical means, as Aristotle called them) such as weeping, wailing, witnessing, etc., as well as from those discursive techniques that are based on the form of presentation or exposition.

Let us now examine the fundamental reason why science should be included in the domain of argumentation. Here Aristotle comes to our aid more than Perelman, who has always been against such an operation. Aristotle considered scientific knowledge (*epistéme*) a (syllogistic) deduction from principles (definitions). This, however, represents only the second segment of scientific method (procedure). The first concerns our knowledge of the principles. How do we arrive at them?

Since scientific knowledge is expressed by propositions in which a predicate is attributed to an object, and since these propositions must be the conclusion of a syllogism, the problem of attaining principles amounts to finding a middle term between the subject and the predicate of the conclusion.[49] The solution to this problem lies in dialectics.[50] In the *Topics*, Aristotle stated that "it is through *éndoxa* about them [the principles] that these have to be discussed," adding that dialectics is useful for this purpose because "the ability to puzzle on both sides of a subject will make us detect more easily the truth and error about the several points that arise."[51] In the *Rhetoric*, he claimed that "we must be able to employ persuasion, just as deduction can be employed, on opposite sides of the question, not in order that we may in practice employ it in both ways (for we must not make people believe what is wrong), but in order that we may see clearly what the facts are, and that, if another man argues unfairly, we on our part may be able to confute him."[52]

These passages illustrate why Aristotle thought dialectics plays a role in science. The scientist is required to prove propositions — that is, to prove that certain predicates necessarily belong to certain subjects — starting from proper universal principles. As these principles are intuitive (they are grasped by the *nous*) but not immediately available, the scientist, in order to find them, starts off with a set of particulars and begins a process of "induction" (*epagogé*), until he comes to "see" and leads his interlocutors to "see" the principles by an act of final intuition. As these particulars are observations in the double sense of observed things (*phainomena*) and widely accepted reports on such things (*legomena*),[53] this means that the scientist arrives at universal principles by collecting data and opinions, discussing rival views, and convincing his interlocutors that one view is better than another.[54] This is why dialectics, which is the art of confutation, and rhetoric, which is the art of persuasion, have a role to play in science. They constitute the "logic of scientific discovery."[55]

If we now join this first part of method (procedure) with the second, a typical sequence of steps in Aristotle's view of scientific inquiry would be:

(A) $O \ldots P \ldots C$

Where O is *phainomena* in the double sense mentioned above, P is principles, and C the cognitive claim.

With the birth of modern science, this procedure underwent several transformations. The ideal of the Aristotelean *epistéme* was not rejected but confined to specific areas, in particular that of applied mathematics. Here, as Galileo wrote in his early work *Mechanics,* after collecting observational, data, we first introduce "definitions of terms" and "first suppositions" (or "axioms"); then we derive "true demonstrations. . .as from fertile seeds."[56] Thus Galileo transformed procedure (A) into procedure:

(B) $O \ldots A \ldots T$

where O stands for observational data (Aristotle's *phainomena* in the first sense), A for certain axioms, and T for a deduced theorem.

In the field of empirical sciences, procedure (A) underwent deeper transformations, because self-evident principles were abandoned in favor of hypotheses. Again Galileo is a useful source. Take, for example, his *Notes* to Antonio Rocco's *Esercitazioni filosofiche* (Philosophical Exercises).[57] In these *Notes,* Galileo describes how he arrived at his law of falling bodies. He started with observations (he writes that he observed "during storms tiny grains of hail falling alongside both middle-sized ones and others as much as ten times bigger, and the bigger ones did not fall on the ground before the others"). He went on to formulate a hypothesis ("from here I formulated an axiom that could not be put in doubt by anyone, and I supposed that any heavy falling body had in its movement a rate of velocity limited and predestined by nature"). He then proceeded to make deductions from the hypothesis, using a thought experiment ("once this supposition had been made, I imagined with my mind . . .") and to perform other tests. Finally, he drew his conclusion. Thus, in the field of empirical sciences, procedure (A) was transformed into procedure:

(C) $O_i \ldots H_p \ldots O_t \ldots C_t$

that is the hypothetico-deductive procedure we derived from another passage from Galileo in chapter 1.

Clearly, procedures (B) and (C) differ from (A). And yet the reasons that induced Aristotle to find a place for dialectics and rhetoric in procedure (A) are equally present in (B) and (C). Consider a few problems concerning (C) for now. Given certain phenomena, several explanatory hypotheses can be introduced with the same observational consequences; two hypotheses can be observationally equivalent; if one hypothesis explains a phenomenon better than another, the other may nonetheless have other advantages; in order to be taken as a candidate for explanation, a hypothesis must have some initial plausibility; furthermore, a test must be "severe," an observation must be "reliable," an experiment must be "properly conducted." When problems like these are to be solved, *decisions* must be made and argued: what can we resort to if not persuasive arguments, that is, rhetoric?

A methodologist might object that the role we are giving rhetoric is already played by inductive logic and methodology. This is precisely the point we are debating. We have seen that methodological rules have an open texture that can be tightened only through decisions that have to be well-argued. But making decisions and arguing for them involves discussing rival views and convincing an audience. This is the fundamental reason why rhetoric enters into science. A deeper analysis of the idea of a scientific code will allow us to spell it out.

2.4 Reasons for Rhetoric in Science.

In the previous section I compared a scientific code to a set of recommendations given to a prudent driver. A more illuminating and useful comparison is a legal code (but a motor vehicle code would do as well). Positive and negative analogies between a scientist and a judge illustrate why rhetoric plays an essential role in science.

The first reason concerns the *application* of the code. Take a judge who is called upon to adjudicate a case where an individual is accused of having committed a deed *D*. The judge will reach a verdict by subsuming *D* under a law *L* of the legal code. The argument leading to his verdict will be a practical syllogism of the following kind:

Anyone who commits *D* must be punished with *P;*
a has committed a *D*-like deed *x;*
∴ *a* must be punished with *P.*

A syllogism such as this raises two questions: one regards the minor premise, the *quaestio facti;* the other, the major premise and the connection between the two, the *quaestio juris.* Let us start with the former.

In order to reach a verdict, a judge must deal with a number of preliminary but fundamental decisions. He must, for example, make sure that *a* has really committed *x,* and that *x* is legally relevant. He must also decide whether *x* is *D*-like before establishing whether *a* is liable for punishment. He must, moreover, define precisely which law *L, D* falls under before deciding what punishment to give to *a,* and he must judge whether *D* is a serious violation of *L* before sentencing *a* fairly, and so on. Only after making these decisions will the judge be able to consider the *quaestio juris* and construe the syllogism. What arguments will he use for adopting the needed decisions? Since they cannot be arbitrary, he will have to provide "good reasons" for them. He will obviously rely on legal precedents, establishing similarities and differences, considering the general principles of the legal code he is called on to apply, etc. In any case, he will form an opinion and he will justify it with a series of nonsyllogistic arguments—e.g., arguments *a pari, a contrario, a fortiori, a majori, ad minus,* etc.—which make up the *juridical argumentation (or rhetoric).*

A scientist is in an identical position. Let us take an abstract yet typical situation.

Suppose a given observational consequence *O* is derived from a theory *T* and that *O* actually takes place, say, during an experiment. A scientist will reach a verdict on *T* by subsuming this situation under an appropriate rule of his scientific code. His argument might be a practical syllogism of the following kind:

Theories confirmed by experiments *O* are to be accepted;
T is confirmed by an *O*-like experiment *e;*
∴ *T* is to be accepted.

A syllogism such as this raises the same two kinds of problems as our judge's syllogism did. Let us consider the first kind.

Here too some preliminary but fundamental decisions concerning the *quaestio facti* are necessary. A scientist, for example, must ascertain whether *e* is *O*-like, i.e. whether it is a reliable experiment, in order to legitimize his verdict on *T.* He must also decide whether *e* is a severe test or not, in order to establish what fair treatment *T* should undergo. For the same reason he must, moreover, decide whether rule R of his code (major premise) is pertinent to the T-e relationship (minor premise). Suppose e is judged reliable and severe, but T is considered to have some important anomalies, say e', that are explained by a rival theory T'; or suppose e is reliable and severe, T is free of important anomalies but happens to conflict with other accepted theories or with dominant philosophical or metaphysical assumptions: the scientist will then have to decide whether a given rule R is pertinent to his case or whether it might not be better to apply a different rule R'. What kinds of arguments will he use to justify his decisions? He will clearly use ***deduction*** when he infers e from T; this inference is regulated by the rules of formal logic and mathematics. He will also use ***induction*** when, for example, he states that T is probable given e, and when, again, he claims that since e was a priori (that is, before the experiment) fairly improbable, then T is a posteriori (that is, after the experiment) all the more probable. These inferences are regulated (if at all) by the canons of inductive logic, and they are sometimes formalized by theorems of probability calculus (e.g., Bayes's theorem). However, deductive and inductive inferences are useless to a scientist when he has to make those preliminary decisions which lead him to delivering a verdict on T. Such inferences come *afterwards*. The arguments a scientist has to use to declare e reliable or strict, e' of little importance, T initially probable, or R more (or less) pertinent than R' to the case in question, etc. are neither deductive nor inductive. Inasmuch as the scientist will have to refer to considerations of opportunity and value in order to convince his interlocutors, his arguments will be *rhetorical arguments*.

Methodologists neglect this point either because they do not take the *quaestio facti* seriously or because they leave it aside altogether. Indeed, by its very nature, methodology comes to the fore when such problems have been answered, that is, when there is a well-formulated hypothesis or theory on one side, and hard facts on the other. Methodology functions like a recipe that starts with "dice the turkey breast" without mentioning how one is supposed to get the turkey; that is, without considering how to introduce a good theory and how to make facts talk.

Naturally this omission is not by chance. It stems from the fact that the methodological approach is often associated with the idea that theories are invented, not inferred, and it is more often based on the empiricist dogma that facts can be obtained with little trouble thanks to observation and experiments. Encouraged by this double conviction, the methodological approach has felt free to devote itself uniquely to the rules that govern the relationship between hypothesis or theory on the one hand, and facts on the other. Even when it took more than one factor into consideration—

such as when background knowledge was added to theory, or when theory was replaced by a series of theories—facts remained "the 'impartial arbiter' of scientific controversy."[58]

Clearly, if the stage is vacated, leaving only two protagonists—facts and theories—the typical relationship between them would only be deductive or inductive. Equally clearly, though, if one realizes that facts rarely express themselves unequivocally, especially in the most delicate circumstances, and thus if one brings back onto the stage, alongside facts and theories, an interpreter who allows the former to communicate with the latter, then one can no longer deny that the interpretation and selection of facts and the mediation that goes on between facts and theories must be taken into account before analyzing their relationships into chains of deductive or inductive arguments.

The second reason why rhetoric plays an essential role in science has to do with the *quaestio juris,* in particular with the *interpretation* of the rules of the scientific code. Here too the judge-scientist analogy is illuminating.

Suppose our judge made all the necessary preliminary decisions, establishing that x is D-like and that D falls under L. The problems concerning the laws L of the code—that is the major premise of the practical syllogism—now come in. In order to clarify these problems, it will be useful to remember, as Perelman has stressed,[59] that a legal code has none of the characteristics of a formal system, not even an applied formal system such as a game.

First, a legal code is often *vague,* because even the most detailed rule often has fuzzy edges. Consider, for example, a warning outside a library that reads "Noise is forbidden." This evidently means it is forbidden to sing or play an instrument in the reading room; but does it also mean one is not allowed to shout out in the event of danger? In the abstract, a code could of course discipline even situations of this kind in detail, making the meaning of each term used as precise as possible; but no code will ever be able to eliminate altogether the fuzziness of the terms in the natural language in which it is formulated. As Perelman commented, "in law, the use of vague notions is not always a defect."[60] To the contrary, a literal interpretation of a very detailed rule is often a defect: *summum jus, summum iniuria,* as the saying goes.

Second, a legal code is *incomplete,* it inevitably has lacunae because it simply cannot discipline all the cases under its jurisdiction. Usually, codes make up for this defect by introducing phrases such as "case of superior force," "the invincible force of events," "extraordinary situation," and so on, that are left to the discretional interpretation of the judge.[61] This is not considered a defect, though: the judge's discretion allows equity and justice to work hand in hand.

Third and finally, a legal code is often *antinomic:* there are situations where two rules prescribe contradictory actions because an earlier decision overlapped with another legislator's later decision, or because the same case may be subject to two different sets of legal regulations.[62]

In short, putting all these characteristics together, we could say that even the most precise, detailed legal code always has an "open texture."

If these defects cannot be eliminated, one wonders how a judge manages to make the major premise of his syllogism precise enough to reach a fair verdict. The answer is that he can do so only by *interpreting* the rules in his code. Just reading the regulations is hardly ever enough. Far from it: a judge must compare a given rule to other pertinent ones, and compare the sum of rules to the set of values the code is supposed to protect. What kind of arguments will he use? Normally, rhetorical ones.

Here again, the circumstances are surprisingly similar to those in science. To start with, it is easy to show that even the best scientific codes present the same characteristics as legal codes. Since "scientific code" here has its usual meaning of a set of methodological rules, let us stick to these rules. My goal is to demonstrate that a code in that sense does not exist, to deconstruct the idea, as it were, from within.

Vagueness. The rules of science always require a scientist to make personal decisions about their interpretation. In most cases, this presents a margin of vagueness.[63] What, for instance, does "consolidated" mean in the rule: "Reject any hypotheses disproved by *consolidated* observational data"? One can always make the effort to be more precise by formulating definitions, but there is a limit to precision, and who is to say that eliminating vagueness is necessarily a good thing? The saying, *summum jus, summa iniuria,* also goes for science. As we saw in chapter 1, it was the elasticity of interpretation of the rejection rule that allowed Galileo to save the Copernican system from disconfirmation.

Incompleteness. In order to make a decision concerning a cognitive claim or a course of action, a scientist not only has to choose a pertinent rule and interpret it, he must also establish the limits within which it can be applied. A rule such as "Do not use ad hoc hypotheses" follows the same pattern as the rule "Do not disturb the neighbors": it prohibits some things but cannot specify all of them. Methodological rules typically contain an implicit "unless-clause": "Do not accept uncontrolled theories *unless* x"; "Reject theories disproved by observations *unless* y"; "Select a theory with greater empirical content *unless* z," where x, y, and z are variables that only the practicing scientist can ponder within a debate with his interlocutors. One might object that substituting these variables with constants that refer to well-defined circumstances of applications is precisely the goal of methodology; it is practically impossible, however, to foresee all possible circumstances for a rule. Moreover, to do so would not necessarily be desirable because incompleteness, like vagueness, is not necessarily a bad thing.

Antinomy. There are situations in which a decision can fall in between two domains with two different sets of rules that contradict each other. Take, by way of example, the acceptance rule we drew from Galileo in chapter 1, that is,

AR.3 Only those hypotheses that are falsifiable by observational data are to be accepted.

In Popper's interpretation, this forbids accepting hypotheses that utilize immunizing devices. On the other hand, the rejection rule that saved Galileo's research, that is,

RR.3 Any hypothesis whose observational consequences are contradicted by consolidated observational data, unless they constitute a local or secondary anomaly, is to be rejected,

allows one to maintain theories that are not seriously disconfirmed. Let us suppose, then, that Galileo had considered a certain fact, say the eccentricity of the orbits of Venus and Mars, an anomaly of Copernicus' theory. Should he have accepted the theory and adjusted it ad hoc to make it compatible with RR.3, or should he have rejected it because this adjustment would violate AR.3?

If a scientific code has the same lacunae as a legal code, the same question applies: how can a scientist patch up these holes before reaching a verdict? The answer is always the same—by *interpreting* the code. Like a judge, a scientist will base this interpretation on his education and culture, taking into consideration precedents, the meaning he attributes to the general values that inspire his code, the consequences of his decisions, the value of other codes he considers better, etc. Like a judge, a scientist will decide after discussing with others who hold rival interpretations and trying to convince them that his reasons are better. Like a judge, a scientist will defend his interpretation by attacking, confuting, and persuading, i.e., using rhetorical arguments.

There is a third and final reason why rhetoric enters into science: it has to do with the *choice* of a scientific code. A negative analogy between a scientist and a judge is illuminating here. Whereas a judge can only respect the values of the legal regulations he works with, a scientist can raise questions about the values inherent in the scientific code of the day, and he can also change the rules. While a judge is executor, a scientist is both executor and legislator because he can propose new values or different hierarchies for the accepted rules; he can uphold the jurisprudence he has learned during his training, but he can also change and create a new precedent for the future. When Galileo replaced geocentrism with heliocentrism, he was not just changing an old theory, he was also changing at least part of the current scientific code. When Darwin challenged fixed species in favor of evolution, he replaced Bacon's standards with the hypothetico-deductive method. When Bondi and Gold advanced the steady-state theory of the universe, they also raised doubts about the inductive and empirical method typical of their discipline and went on to opt for the deductive method. As we have already remarked, changes in theory are often associated with changes in rules. And changes in rules are never justified by other rules but by the fact that supporters of change manage to use arguments that are stronger and more convincing than those of their adversaries. Moreover, a scientist is often a defense lawyer rather than a trial judge: believing his cause to be just, he will use all the persuasive arguments at his disposal, from the noblest to the most question-begging, in order to achieve his goal.

Before concluding this chapter, an important warning is needed concerning the path I intend to follow. I propose rejecting all three theses in the Cartesian project and denying the existence of a scientific code, at least in the technical sense of a set of precise rules. I do not, however, propose either abolishing the constraints of science by saying that "anything goes," or assimilating a scientific discussion to a political or ethical conversation. Rhetoric does not necessarily lead to these two conclusions: neither to the former because rhetoric *does have* constraints; nor to the latter because scientific rhetoric has *specific* constraints. Once again the analogy between a scientist and a judge is illuminating.

Even though, as Perelman says, a legal text underdetermines his decisions, a judge is not free to act as he thinks best; nor are all his decisions equally admissible. When the interpretation of a norm is made by individuals all of whom accept the legal regulations and have the same training, the verdicts reached, in most cases, are sufficiently univocal.[64] The same is true for a scientist. Even though, as Kuhn maintains, scientists sharing the same values "may nevertheless make different choices in the same concrete situation,"[65] in most cases, trained as they are according to the same values, they draw the same conclusions. Values are constraints. Some interpretations and breaches of these constraints are compatible with the physiological imperfection of any code, but others are hardly tolerable and, when subjected to critical debate, are clearly untenable, even for the most flexible jurisprudence.

Take acupuncture as an example, and suppose that Françoise and the Guru—to return to their dialogue in chapter 1—agreed on the fact that testability through experiments is an essential value for scientific inquiry. If this were the case, Françoise would find it easy to object that all the hypotheses about two kinds of energy, about perverse energy, energy imbalances, meridians that channel energy, etc., are *not* empirically testable at the present stage of scientific culture and jurisprudence, under any interpretation that was not deliberately captious about the term "testable." A verdict that excluded acupuncture from the realm of science in the name of the value of testability would thus be a sufficiently univocal verdict.

All this, of course, does not provide a solution. We are, rather, at the threshold of a philosophical research program. From now on, I shall reserve the term *scientific rhetoric* for those persuasive forms of reasoning or argumentation that aim at changing the belief system of an audience in scientific debates, and the term *scientific dialectics* for the logic or canon of validation of those forms. In the case of law, a juridical logic already exists that studies specifically the validation of juridical arguments.[66] In the case of science, a deductive logic and, although more precariously, fragments of inductive logic, exist, but we still know very little about scientific dialectics. The Cartesian methodological tradition never allowed the concept to be formulated (and smothered the noble old concepts of dialectics and rhetoric). And yet, if we admit the failure of the Cartesian project and do not want to give up trying to understand how scientific inquiry actually proceeds, we have to explore this logic.

The difficulty of the enterprise suggests that I advance step by step. We shall start by examining a sample of the multifarious persuasive arguments scientists use and the different roles they play in various contexts. My first goal is to become acquainted with scientific rhetoric, its techniques and its functions. Only then shall I take up the question of constraints and try to construct a canon for scientific rhetoric, that is, scientific dialectics.

Three

The Rhetoric of Science

A careful study of the reasoning employed by the creative and original thinkers, both in science and in philosophy, would reveal that reasoning is infinitely more varied than anything to be found in the manuals of logic or scientific methodology.

Ch. Perelman, *The Idea of Justice and the Problem of Argument*

3.1 Galileo's "Flowers of Rhetoric"

The main aim of Galileo's *Dialogue Concerning the Two Chief World Systems,* although political prudence obliged him to be cautious, was to prove that Copernicus' system was better than Ptolemy's. To achieve this aim, Galileo had to justify both his method of proof and the specific theories advanced with it. The task was titanic. To justify his method, Galileo had to elaborate a theory of the sources of knowledge, of truth and error—in short an epistemology and an anthropology. His theories required a new dynamics, a new optics, a new cosmology. Feyerabend was right when he wrote: "what is needed for a test of Copernicus is an entirely new world view containing a new view of man and his capacities of knowledge."[1] Was he also right, however, when he commented that, in Galileo's circumstances, "allegiance to the new ideas will have to be brought about by means other than arguments. It will have to be brought about by irrational means such as propaganda, emotion, *ad hoc* hypotheses, and appeal to prejudices of all kinds"?[2] Let us see how Galileo proceeded, starting with the justification of his method of "sensory experiences and necessary demonstrations."

The expression is rather vague. Positively, it means that science can only accept conclusions proved by mathematical demonstrations and based on direct observations or experiments actually performed. Negatively, it means that in science, as Galileo once objects to Simplicio, one must not "entangle these little flowers of rhetoric in the rigor of demonstrations."[3] Galileo lets us know the reasons why he condemns rhetoric: rhetoric is the art of oratory, and "in the natural sciences the art of oratory is ineffective" because when one has to demonstrate "true and necessary" conclusions, "a thousand Demosthenes and a thousand Aristotles would be left in the lurch by every mediocre wit who happened to hit upon the truth for himself."[4] This view is typical of the fathers of modern science. Like Bacon, Descartes, and many others, Galileo maintained that science is not a dialogue between various interlocutors (even less a dialogue with one interlocutor only, say Aristotle), but a dialogue with nature. Since we have access to nature through our senses, and since it is assumed that nature's basic structure is mathematical, a dialogue with nature must be conducted through "sensory experiences and necessary demonstrations." But this very argument is enough to show that Galileo's condemnation of rhetoric is exaggerated: for how can it be proved that the only (or best) access to nature is direct observation? And what arguments can be used to demonstrate that the book of nature is written in mathematical characters? These arguments can evidently be neither mathematical nor empirical. Obviously, they must be rational, but in what way?

Let us see how Galileo proceeded. In his *Dialogue,* each term of the binomial "sensory experiences and necessary demonstrations" comes up against a wall of resistance. In order to achieve his goal, he has to overcome this resistance, and in order to overcome this resistance, he has little choice but to use rhetorical arguments. Consider a typical argument in favor of "sensory experiences."[5]

(1) *Argument by retort.* From the start, Simplicio admits that Aristotle "held in his philosophizing that sensible experiments were to be preferred above any argument built by human ingenuity."[6] He is right because Aristotle had often criticized those predecessors who, rather than saving phenomena by creating theories, preferred to save their own theories by adapting the phenomena.[7] Nonetheless, Simplicio is suspicious of experiences whenever they conflict with Aristotle's conclusions, especially if these conclusions are drawn from apparently intuitive or well-founded premises. This raises the question whether the senses should be preferred to reason. Galileo's typical (though not unique) answer is to rely on the former. How does he argue for it? Let us consider this question in the context of a debate that takes place on the First Day of the *Dialogue* about the alterability of the heavens.

(1.1)

Salv. Whenever you wish to reconcile what your senses show you with the soundest teachings of Aristotle, you will have no trouble at all. Does not Aristotle say that because of the great distance, celestial matters cannot be treated very definitely?

Simp. He does say so, quite clearly.

Salv. Does he not also declare that what sensible experience shows ought to be preferred over any argument, even one that seems to be extremely well founded? And does not he say this positively and without a bit of hesitation?

Simp. He does.

Salv. Then of the two propositions, both of them Aristotelian doctrines, the second—which says it is necessary to prefer senses over arguments—is a more solid and definite doctrine than the other, which holds the heavens to be inalterable. Therefore it is better Aristotelian philosophy to say, "Heaven is alterable because my senses tell me so," than to say "Heaven is inalterable because Aristotle was so persuaded by reasoning."[8]

This kind of argument, repeated throughout the *Dialogue,*[9] is an argument by retort. Galileo tries to demonstrate that Simplicio is rejecting a conclusion based on a criterion of proof that implies it. He therefore reminds him of the constraint of coherence, and presents him with a dilemma: either the criterion is acceptable and so is the theory of alterability, or the theory of inalterability is acceptable and thus Aristotle disobeys his own criterion. Galileo's argument does not positively prove the priority of the senses over reason, but it tends to persuade his interlocutor that this criterion is valid because the interlocutor (or the authority that he recognizes) professes it.

This is the essence of scientific argumentation: it tries to induce agreement in the interlocutor on the basis of certain assumptions he either accepts or is tied to. An argument of this kind is typically *ad hominem* in the old Aristotelian sense, for it utilizes premises shared by the interlocutor to demonstrate disagreement with a thesis or practice he himself professes.

One might well ask, however, whether there isn't a trick, an attempt to disarm the adversary with illegitimate weapons? *Yes,* since what Galileo tries to pass off as a "better basis" for demonstrating the alterability of the heavens is not properly speaking "sensory experience" but what we might call "instrumental experience," namely, telescopic observations. *No,* since Galileo can use other arguments to show that his "instrumental experience" is, to a certain extent, just as reliable; and that mere "sensory experience" is often acritical (for example, it is particularly subject to illusion). *Yes,* since in shifting from one experience to another, Galileo also shifts his ground. *No,* because he offers arguments to support this shift, and arguing is not the same as playing tricks.

An examination of Galileo's techniques for persuading Simplicio of the validity of the second term of the binomial, that is, "necessary demonstrations," leads to the same conclusion.

(2) *Argument by counter-example.* The objection raised by Simplicio about Galileo's "necessary demonstrations" is a matter of principles, and concerns the legitimacy of applying them to the natural world. Consider, for example, the following passage from the First Day. The exchange is about the dimensions of a body, and Salviati, using geometry, concludes there are three because from one point only three perpendicular lines can be drawn. Simplicio agrees, but raises several objections:

(2.1)

> *Salv.* I shall not say that this argument of yours cannot be conclusive. But I still say, with Aristotle, that in physical matters one need not always require a mathematical demonstration.
>
> *Sagr.* Granted, where none is at hand; but when there is one at hand, why do you not wish to use it?[10]

A similar argument is found in the Second Day. Sagredo praises applied mathematics, saying: "The argument is truly very subtle, but nonetheless convincing, and it must be admitted that trying to deal with physical problems without geometry is attempting the impossible." To which Simplicio raises the usual objection: "mathematics may prove well enough in theory that *sphaera tangit planum in puncto,* a proposition similar to the one at hand; but when it comes to matter, things happen otherwise."[11] Salviati then argues that Simplicio is wrong, maintaining that "material spheres are subject to many accidents to which immaterial spheres are not subjected," which "prevents things taken concretely from corresponding to those considered in the abstract."[12] Salviati's rejoinder is the following:

(2.2)

> *Salv.* Then, whenever you apply a material sphere to a material plane in the concrete, you apply a sphere which is not perfect to a plane which is not perfect, and you say that these do not touch each other in one point. But I tell you that even in the abstract, an immaterial sphere which is not a perfect sphere can touch an immaterial plane which is not perfectly flat in one point, but over a part of its surface, so that what happens in the concrete up to this point hap-

> pens the same way in the abstract. . . . Do you know what does happen, Simplicio? Just as the computer who wants his calculations to deal with sugar, silk, and wool must discount the boxes, bales, and other packing, so the mathematical scientist, when he wants to recognize in the concrete the effects which he has proved in the abstract, must deduct the material hindrances, and if he is able to do so, I assure you that things are in no less agreement than mathematical computations.[13]

In both exchanges (2.1) and (2.2), Galileo tries to confute his interlocutor by producing a counter-example. When Simplicio objects that mathematics cannot give rigorous proofs when dealing with nature, Galileo counters with two examples. Especially in (2.2), the counter-example is derived from what Simplicio himself admits during the discussion. Nevertheless, Galileo's argument does not prove his thesis as conclusively as he had hoped. Simplicio's objection is that mathematics cannot be applied to nature in the sense of "sensible and physical matter"; Galileo's answer is that mathematics can be applied to nature when you "deduct the material hindrances," that is, nature in the sense of *ideal, abstract matter,* or, as Galileo says in his *Discourses,* matter which is "perfect and inalterable and free from all accidental change."[14] In (2.2), therefore, we have a *metabasis eis allo ghenos* as in (1.1), with the aggravation that in (1.1) Galileo shifts from sensory experience to instrumental experience, while in (2.2), the shift is much greater, from sensory to *thought* experience. Is there not a trick?

Once again the answer is as much *yes* as it is *no. Yes,* since Galileo's answer is not strictly pertinent; *no,* since Galileo again uses arguments to support his answer, in particular the analogy "that what happens in the concrete up to this point happens the same way in the abstract."[15] *Yes,* since instead of necessary demonstrations Galileo offers rhetorical arguments; *no,* because rhetorical arguments are a perfectly rational means of persuasion. Only if one takes Galileo's condemnation of rhetoric literally can one conclude that the conversion he wants to bring about in his interlocutors takes place through "means other than arguments" or "irrational means." But Galileo cannot be taken literally: when he is in the position of having to justify his own method, he is the first to rely on the very same "flowers of rhetoric" he condemned.

And there are more flowers to come. Even if we take "sensory experiences and necessary demonstrations" in the liberal sense examined earlier, is it really true that the superiority of the Copernican system over the Ptolemaic one can be proved only on these grounds? The answer is decidedly negative. It is not just that the Copernican theory is no better than the Ptolemaic as far as empirical consequences are concerned. Although it explains new facts (such as the phases of Venus), several deeply-rooted assumptions have to be changed for the new theory to be accepted, for example, the distinction between heavenly and earthly regions, the inalterability of the heavens, the nonoperativity of relative motion, etc., which cannot by their very nature be proved empirically or mathematically. Thus, Galileo is obliged to use "flowers of rhetoric" in questions of content, not only in matters of method.

An exhaustive analysis of these flowers is not my goal. The examples that follow are merely a representative sample to show that rhetoric plays an essential role in Galileo's *Dialogue*.[16]

(3) *Argument of parts and the whole.* At one point in the First day, Aristotle's "sensory experiences" are mentioned in support of the correspondence between celestial bodies and their motions. The thesis put forward by Simplicio is that straight motion is proper only to earthly bodies. Salviati raises the following objection:

(3.1)

> Now just as all the parts of the earth mutually cooperate to form its whole, from which it follows that they have equal tendencies to come together in order to unite in the best possible way and adapt themselves by taking a special spherical shape, why may we not believe that the sun, moon, and other world bodies are also round in shape merely by a concordant instinct and natural tendency of all their component parts? If at any time one of these parts were forcibly separated from its whole, is it not reasonable to believe that it would return spontaneously and by natural tendency? And in this manner we should conclude that straight motion is equally suitable to all world bodies.[17]

Salviati wants to convince Simplicio that what is true of the parts is also true of the whole. The argument is effective because it is based on a principle introduced by Simplicio himself, that *eadem est ratio totius et partium.*

(4) *Ad hominem argument.* The debate concerns whether heavenly bodies are generable and corruptible. Simplicio claims they are not and sets out to prove it by invoking the Aristotelian argument that if there is generation and corruption in the heavens, there must also be contrariety. To which Salviati replies:

(4.1)

> The original source from which you derive the contrarieties of the elements is the contrariety of their motions upward and downward.

> Therefore it must be that whatever principles those motions depend upon are likewise contrary to each other. Now since whatever moves upward does so because of lightness, and whatever downward does so because of heaviness, lightness and heaviness must be contrary to each other. No less ought we to consider as contraries any other principles that are the cause of one thing being heavy and another light. According to you yourself, levity and gravity occur in consequences of rarity and density; therefore density and rarity will be contraries. Now these qualities are to be found so abundantly in celestial bodies that you deem the stars to be merely denser parts of their heaven. . . . There being, then, such contrariety between celestial bodies, they must necessarily be generable and corruptible in the same way that elemental bodies are, or else contrariety is not the cause of corruptibility, etc.[18]

This argument is *ad hominem*. It does not prove—or intend to prove—the thesis of the corruptibility of heavenly bodies; all it does is down the adversary with his own arms by showing that contrary conclusions can be drawn from his own premises. It is a successful move on Salviati's part because Simplicio, having been wrecked "in a boundless sea from which there is no getting out," is forced to abandon his line of argument and start thinking in terms of empirical evidence.

(5) *Argument ad personam.* Salviati considers Simplicio's naked-eye observations unreliable, citing as examples comets, new stars, and the most recent discovery of sunspots observed with the telescope. The debate is heated because Simplicio does not accept Salviati's interpretations of these phenomena. In the end, admitting his inability to convince Simplicio, Salviati withdraws, leaving Sagredo to make this comment:

(5.1)

> Those who so greatly exalt incorruptibility, inalterability, etc. are reduced to talking this way, I believe, by their great desire to go on living, and by the terror they have of death. They do not reflect that if men were immortal, they themselves would never have come into this world. Such men really deserve to encounter a Medusa's head which would transmute them into statues of jasper or of diamond, and thus make them more perfect than they are.[19]

This argument proves nothing at all, and would seem irrelevant to the subject under discussion. It is a good example of an (*ante litteram*) psychoanalytical move whose only purpose is to cast doubt on the adversary's personality by pointing out the hidden roots of the arguments he puts forward. Instead of the "manifest" content of Simplicio's thesis, what is criticized here is its "latent" origin.

(6) *Argument by comparison.* Psychoanalysis does not put Simplicio off. He rebuts the assertion that, if one considers function, an earth made of rock would obviously be imperfect; not so heavenly bodies, though, whose only purpose is to serve the earth. Sagredo than changes tack:

(6.1)

> Besides, it seems to me that at such times as the celestial bodies are contributing to the generations and alterations on the earth, they too must be alterable. Otherwise, I do not see how the influence of the moon or sun in causing generations on the earth would differ from placing a marble statue beside a woman and expecting children from such a union.[20]

This argument is more pertinent than the previous one. Sagredo tries to show that Simplicio is attempting to establish a principle (that if A is needed for B, then A may have a different nature from B) that is untenable because it is subject to a very familiar counter-example.

Another comparison can be found in Salviati and Sagredo's joint answer to Chiaramonti's objections, reported by Simplicio, that since the earth is a corruptible body it cannot be in perpetual motion. Such is the case of animals that move about naturally and get tired:

(6.2)

> Then as to the objection against the perpetual motion of the earth, taken from the impossibility of its keeping on without being fatigued, since animals themselves that move naturally and from an internal principle get tired and have need of repose to relax their members. . . . It seems to me that I heard Kepler answering him that there are also animals which refresh themselves from weariness by rolling on the ground, and that hence there is no need to fear that the earth will tire; it may even reasonably be said that it enjoys a perpetual and tranquil repose by keeping itself in an eternal rolling about.[21]

Salviati is the first to admit that this argument is a "joke." Nevertheless, it is effective because it starts with a comparison introduced by Simplicio himself and provides another familiar comparison as a counter-example.

(7) *Argument based on easiness.* The second of the "reasons which seem to favor the earth's motion" gives rise to a discussion on whether circular motions are contrary to each other or not. Sagredo brings the debate to an abrupt end:

(7.1)

> "Contrary" or "not contrary," these are quibbles about words, but I know that with facts it is a much simpler and more natural thing to keep everything with a single motion than to introduce two, whether one wants to call them contrary or opposite. But I do not assume the introduction of two to be impossible, nor do I pretend to draw a necessary proof from this; merely a greater probability.[22]

A similar argument is used successively regarding the alleged motion of the fixed stars which is the sixth reason "which seems to favor the earth's motion":

(7.2)

> It seems to me that it is much more effective and convenient to make them immovable than to have them roam around, as it is easier to count the myriad tiles set in a courtyard than to number the troop of children running around them.[23]

These arguments do not prove the mobility of the earth or the immobility of the fixed stars, at least not in the sense that they adduce empirical evidence. They intend to create consensus by showing that such theses are the simplest solutions of all the ones that are available. This consensus is supposedly guaranteed by the fact that the greater utility of the thesis is the result of certain values (such as economy and clarity), of certain principles (such as *natura nihil frustra facit* and *frustra fit per plura quod potest fieri per pauciora*), that are generally accepted and, more important, are shared by the interlocutor. It is interesting to note that Galileo himself, commenting on (7.1), admits that he is not trying to "draw necessary proof" from his arguments, but simply "a greater probability."

(8) *Argument from a model.* At the end of the Second Day, Galileo examines Chiaramonti's objections to Kepler's argument that since "it is harder to stretch the property beyond the model of the thing than to augment the thing without the property," it is more reasonable to follow Copernicus, who increased the stars' orbits without awarding them motion, than Ptolemy, who increased the velocity of the fixed stars. Salviati replies that Chiaramonti misunderstood Kepler, and that Kepler's reasoning is conclusive:

(8.1)

> If the author's reply is to have any bearing upon Kepler's argument, this author will have to believe that it is all the same to the motive principle whether a very tiny or an immense body is moved

> for the same time, the increase of velocity being a direct consequence of the increase in size. But this is contrary to the architectonic role of nature as observed in the model of the smaller spheres, just as we see in the planets (and most palpably in the satellites of Jupiter) that the smaller orbs revolve in the shorter times. For this reason Saturn's time of revolution is longer than the period of any lesser orb, being thirty years. Now to pass from this to a much larger sphere, and make that revolve in twenty-four hours can truly be said to go beyond the rule of the model.[24]

This argument tries to prove that the order of the planets and the Medicean stars (the relation between position and speed) constitutes a model that should also hold for the fixed stars. The basis of the argument—which could also be considered an argument of reciprocity or by analogy—is the symmetry of nature, which would be thrown into disarray if one accepted Ptolemy's theory.

(9) *Pragmatic arguments.* One of the strongest objections to the Copernican system is that no parallax from the motion of the earth has been observed. Salviati first replies that this is "both on account of the imperfection of astronomical instruments, which are subject to much variation, and because of the shortcomings of those who handle them with less care than is required."[25] He then counterattacks by recalling that "Ptolemy distrusted an armillary instrument constructed by Archimedes himself for determining the entry of the sun into the equinox." Finally, he promises new and more accurate observations, and concludes:

(9.1)

> And if in the course of these operations any such variation shall happen to become known, how great an achievement will be made in astronomy! For by this means, besides ascertaining the annual motion, we shall be able to gain a knowledge of the size and distance of the same star.[26]

This argument recommends a line of inquiry because of potential advantages to be gained. In practice, it is an invitation to set aside the present difficulties in the hope of future benefits. Although Galileo's reference to Ptolemy (a real retort) tries to keep us from suspecting an ad hoc maneuver, the argument is an escape route. Inviting someone who already has serious doubts about a theory to hope for future results that will definitely prove it, is tantamount to asking him to make an act of faith. For our purposes, the argument is interesting because it shows how a rhetorical move can be used to escape the strict dictates of a methodological rule (in this case, the rule that obliges one to reject a theory whose observational consequences have not been verified) and to substitute for it a more tolerant rule (in this case, the rule that allows one to save a theory by adding auxiliary hypotheses that have not yet been tested).

(10) *Argument by double hierarchy.* Towards the end of the Third Day, Salviati uses the properties of a lodestone to illustrate how the earth could have several motions and to explain why it always points towards the same part of the sky. One of his arguments is the following:

(10.1)

> And if every tiny particle of such stone has in it such a force, who can doubt that the same force resides to a still higher degree within the whole of this terrene globe, which abounds in this material? Or that perhaps the globe itself is, as to its internal and primary substance, nothing but an immense mass of lodestone?[27]

This argument can be interpreted in various ways: as an argument about the parts and the whole, an argument by analogy, or an argument by double hierarchy. In this last sense, the argument tries to show that since the earth has the same properties as a lodestone, the fact that its axis always points in the same direction is related to how much bigger and better it is.

(11) *Absurdity and ridicule.* Galileo uses this technique several times. A few examples are given below. Simplicio claims the matter of heavenly bodies is solid and impenetrable, and Sagredo replies:

(11.1)

> What excellent stuff, the sky, for anyone who could get hold of it for building a palace! So hard, and yet so transparent![28]

Simplicio calls on Aristotle and the "authority of so many great authors,"[29] but Sagredo catches him out by reminding him of the tale of the man who, confronted with certain anatomical experiments, said:

(11.2)

> "You have made me see this matter so plainly and palpably that if Aristotle's text were not contrary to it, stating clearly that the nerves originate in the heart, I should be forced to admit it to be true."[30]

Simplicio again mentions Aristotle as the first progenitor of the telescope, and Salviati tells him another little tale:

(II.3)

And certain gentlemen still living and active were present when a doctor lecturing in a famous academy, upon hearing the telescope described but not yet having seen it, said that the invention was taken from Aristotle. Having a text fetched, he found a certain place where the reason is given why stars in the sky can be seen during daytime from the bottom of a very deep well. At this point the doctor said: "Here you have the well, which represents the tube; here the gross vapors, from whence the invention of glass lenses is taken; and finally here is the strengthening of the sight by the rays passing through a diaphanous medium which is denser and darker."[31]

When Simplicio invokes the authority of G. Locher, Sagredo is ready to jump down his throat with this remark:

(II.4)

This author must believe that if a dead cat falls out of a window, a live one cannot possibly fall too, since it is not a proper thing for a corpse to share in qualities that are suitable for the living.[32]

Simplicio introduces the notions of sympathy and antipathy as explanations of certain physical phenomena, but Salviati has yet another story up his sleeve:

(II.5)

Now this method of philosophizing seems to me to have a great sympathy with a certain manner of painting used by a friend of mine: "This is where I'll have the fountain, with Diana and her nymphs; here, some greyhounds; there, a hunter with a stag's head. The rest is a field, a forest, and hillocks." He left everything else to be filled in with color by the painter, and with this he was satisfied that he himself had painted the story of Acteon—not having contributed anything of his own except the title.[33]

Let us pause here with our sample of Galileo's scientific argumentation. Before starting, we stressed his faith in the method of "sensory experiences and necessary demonstrations" and his opposition to the "flowers of rhetoric." In the *Assayer*, in a rejoinder to Sarsi, Galileo wrote:

Consider your case carefully, and think that for one who wants to convince someone else of something that is if not false at least highly dubious, it is of great advantage to be able to make use of probable arguments, conjectures, examples, likely comparisons, even sophisms, strengthening one's reasoning and fortifying oneself with texts, the authority of other philosophers, naturalists, rhetoricians, and historians. But to appeal to the severity of geometric demonstrations is very dangerous if one does not know how to handle them well. Therefore, just as there is no objective half-way between truth and falsehood, likewise in necessary demonstrations, either one draws a compelling conclusion or one ends up in inexcusable paralogisms, leaving no room for using limitations, distinctions, distortions of the meanings of words, or other tricks in order to stand on one's own feet; but it is necessary with few words and at the first attack to be either Caesar or nothing.[34]

"Either Caesar or nothing." This is a good way of putting the Cartesian dilemma: either rigorous demonstrations or inconclusive paralogisms.[35] Or, as Feyerabend says, either "well-defined procedures," or "means other than arguments" and "irrational means." There is a difference, however, between Feyerabend and Galileo: Galileo believed that the second horn of the Cartesian dilemma had no right to exist and consequently relied on the first, while Feyerabend is convinced that the first horn of the dilemma is inadmissible and relies on the second. To Galileo, the third thesis of the Cartesian project (if science had no method it would be irrational) would have sounded like a counterfactual statement; for Feyerabend, on the other hand, it is an assertion proved by history. One dilemma brings about another, however: either Galileo was right and he preached what he did not practice, or Galileo was wrong and he practiced what he did not preach. The arguments examined in this section show that the latter is closer to the truth. More precisely, they prove that if Galileo had really kept his word and had not used flowers of rhetoric when dealing with nature, he would never have been able to write his *Dialogue* and the book would have deserved the end it received at the hands of the Inquisition. It is our good fortune that Galileo the scientist did not pay much heed to Galileo the methodologist.[36]

3.2. Darwin's "One Long Argument"

As is well known, Darwin repeatedly wrote that the *The Origin of Species* was "one long argument"; but there has always been considerable controversy over what kind of argument the work presents.[37] Once Darwin wrote that he had worked according to "true Baconian principles,"[38] but he himself recognized that this methodological self-portrait was not very faithful.[39] There are two main interpretations. Some claim that Darwin's method is hypothetico-deductive;[40] others maintain that it is an inference-to-the-best explanation.[41] His work supports both. The first interpretation is supported when he writes, for example, that the struggle for existence "inevitably follows from the high rate at which all organic beings tend to increase,"[42] or "inevitably follows from the high geometrical ratio of increase which is common to all organic beings,"[43] or that many phenomena "follow inevitably from the struggle for life."[44] The second interpretation is supported when he writes that "it can hardly be supposed that a false theory would explain, in so satisfactory a manner as does the theory of natural selection, the several large classes of fact above specified,"[45] or when—contradicting what he had written a few months before in *The Origin of Species,* i.e., that he was "fully convinced of the truth of the views given in this volume"[46]—he admits, in a letter to Hooker dated February 14, 1860, that he had "always looked at the doctrine of Natural Selection as a hypothesis which, if it explains several large classes of facts, would deserve to be ranked as a theory deserving acceptance."[47]

Since both interpretations can be grounded in Darwin's writings, any reconstruction of the *Origin* according to one or the other would necessarily be partial.[48] Darwin, like Galileo, used a *family* of methods. Like anyone else with an important goal to achieve, and a firm belief in that goal, he could not tie his hands behind his back as far as the means to achieve it were concerned. He had, rather, to keep his own options open for choosing the best method case by case. Thus, Darwin chose a complex strategy of argumentation.[49] If we pick only the deductive or inductive arguments out of this strategy, a vital part of his "one long argument" is lost. In fact, Darwin used at least three sets of arguments with distinct aims. In the order in which they appear in the *Origin* they are as follows:

A: arguments of discovery showing that the hypothesis of natural selection is plausible;

B: arguments of defense showing that the hypothesis of natural selection surmounts certain difficulties and objections;

C: arguments of confirmation showing that the hypothesis of natural selection passes the test of facts pertaining to its own original domain and others as well.[50]

It can be shown that the arguments in C are *deductive* and *hypothetico-deductive,* while those in A and B are *inductive* and *rhetorical.* Since the hypothetico-deductive side of Darwin's strategy has been extensively explored, and since what I am interested in here is the role rhetoric plays in science, I shall leave aside the arguments in set C.

Let us begin with the arguments in set A. "Natural selection" is a theoretical concept. Now, one of the most delicate problems facing a scientist who introduces a theoretical entity as a causal agent of observational phenomena is the one described by Newton in his first *Regula Philosophandi:*

"we are to admit no more causes of natural things than such as are both true and sufficient to explain their appearances."[51] To conclude that *H* is confirmed by *O* (or that it is the cause of *O*) it is not enough to prove that *H* implies *O* and to ascertain that *O* is the case; other requisites must be satisfied. Firstly, *H* must be a priori plausible. But how can this plausibility be ensured? Philosophers of science at the time knew of two ways. One was to derive from *H* not only *O* but also other facts *O′* which are different from *O*—that is, different from those facts on which *H* was originally based. The other was to show that *H* is similar to another hypothesis *H′*, already confirmed, that explains a fact *O′* which is similar to *O*. The first way is the "consilience of inductions" recommended by Whewell; the second, the "analogy of causes" suggested by Herschel.[52] Darwin followed both paths in his *Origin*. I shall begin by examining the second.

(1) *Arguments by analogy*. At the beginning of the third chapter of the *Origin*, Darwin establishes his *explanandum* in these terms: "how is it that varieties, which I have called incipient species, become ultimately converted into good and distinct species, which in most cases obviously differ from each other far more than do the varieties of the same species? How do these groups of species, which constitute what are called distinct genera, and which differ from each other more than do the species of the same genus, arise?"[53] We may indicate these facts with

E

taken as the effects to be explained or the results to be obtained from the theoretical explanation. Darwin wrote that "all these results, as we shall fully see in the next chapter, follow inevitably from the struggle of life." This derivation, however, is not direct. More than one step is needed, the first of which is

I

that is, "the high rate at which all organic beings tend to increase."[54] From this high (geometric) rate of increase, with the addition of certain implicit premises, a struggle for existence, *S*, "inevitably follows." As Darwin wrote: "every being, which during its natural lifetime produces several eggs or seeds, must suffer destruction during some period of its life, and during some season or occasional year, otherwise, on the principle of geometric increase, its numbers would quickly become so inordinately great that no country could support the product. Hence, as more individuals are produced than can possibly survive, there must in every case be a struggle for existence, either one individual with another of the same species, or with the individuals of different species, or with the physical conditions of life."[55] Thus, taking Darwin's "inevitably follows" in a deductive sense, the second step is:

$$I \rightarrow S.$$

Now, if there is a struggle for existence, individuals with useful variations will have a greater probability of survival, while those with detrimental variations will tend to disappear. "Owing to this struggle for life, any variation, however slight and from whatever cause proceeding, if it be in any degree profitable to an individual of any species, in its infinitely complex relations to other organic beings and to external nature, will tend to the preservation of that individual, and will generally be inherited by its offspring. The offspring, also, will thus have a better chance of surviving, for, of the many individuals of any species which are periodically born, but a small number can survive. I have called this principle, by which each slight variation, if useful is preserved, by the term of Natural Selection."[56] Thus, taking Darwin's "owing" in a deductive sense, the third step is:

$$S \rightarrow N$$

and the fourth and last step, which is also clearly deductive, is:

$$N \rightarrow E.$$

Up to this point, Darwin's procedure is clearly hypothetico-deductive and we may illustrate it with the diagram below (in which the arrows indicate deductive links):

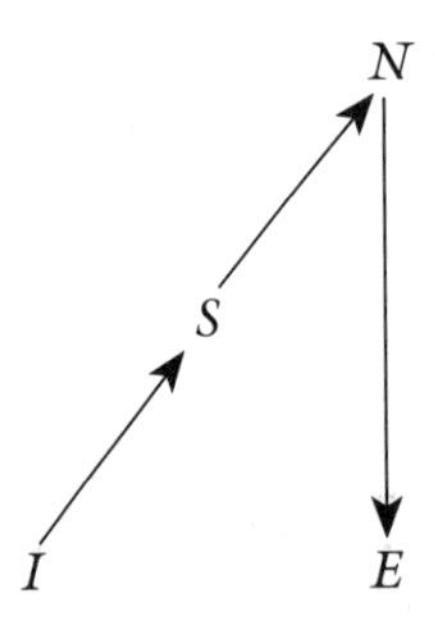

Clearly, the delicate part of this procedure is $S \rightarrow N$. Given that S is observational, N is certainly theoretical. Strictly speaking, then, one cannot say that N follows logically from S. Darwin does not actually say this, but he seems to think it ("*owing* to this struggle"). He is well aware, however, that the problematic nature of $S \rightarrow N$ makes the explanation of E equally problematic, and he realizes that the limits of the hypothetico-deductive procedure lie precisely in this implication. In order to raise the value of *N* to more than simple conjecture, and thus render *N* more acceptable to his interlocutors, Darwin has to find some ***independent*** support in favor of *N* and therefore implement the hypothetico-deductive procedure. Following Herschel, the strategy he adopts is to show that a cause *A* (artificial selection) analogous to *N* implies an effect *D* (domestic variations) analogous to *E* implied by *N*. The implemented procedure is illustrated by the diagram below (in which ≈ indicates the relationship of analogy).

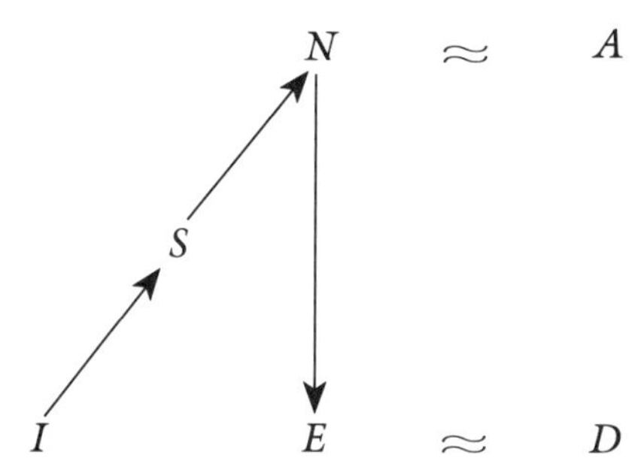

In chapter 4 of the *Origin,* called "Natural Selection," Darwin introduces the *N*-*A* analogy:

(1.1)

Can the principle of selection, which we have seen is so potent in man, apply in nature? I think we shall see that it can act most effectually. Let it be borne in mind in what an endless number of strange peculiarities our domestic productions, and in a lesser degree, those under nature, vary; and how strong the hereditary tendency is. Under domestication, it may be truly said that the whole organization becomes in some degree plastic. Let it be borne in mind how infinitely complex and close-fitting are the mutual relations of all organic beings to each other and to their physical continuations of life. Can it, then, be thought improbable seeing that variations useful to man have undoubtedly occurred, and other variations useful in some way to each being in the great and complex battle of life, should sometimes occur in the course of thousands of generations? If such things occur, can we doubt (remembering that many more individuals are born than can possibly survive) that individuals having any advantage, however slight, over others, would have the best chance of surviving and of procreating their kind? On the other hand, we may feel sure that any variation in the least degree injurious would be rigidly destroyed.[57]

In his last chapter, "Recapitulation and Conclusion," Darwin presents the *E-D* analogy:

(1.2)

> If then we have under nature variability and a powerful agent always ready to react and select, why should we doubt that the variations in any way useful to beings, under their excessively complex relations of life, would be preserved, accumulated, and inherited? Why, if man can by patience select variations more useful to himself, should nature fail in selecting variations useful, under changing conditions of life, to her living products? What limit can be put to this power, acting during long ages and rigidly scrutinizing the whole constitution, structure, and habits of each creature—favouring the good and rejecting the bad? I can see no limit to this power, in slowly and beautifully adapting each form to the most complex relations of life. The theory of natural selection, even if we looked no further than this, seems to me to be in itself probable.[58]

If we now go back to the first way of supporting or reinforcing *N,* the arguments change. Darwin implements the hypothetico-deductive procedure with the "consilience of inductions." When he claims it would not be understandable how a false theory could explain such a great variety of phenomena, meaning by this not only the phenomena of the original *explinandum E,* but also those different facts that are examined in chapters 10–13 of the *Origin*. This new implemented procedure can be illustrated with the next diagram, where E_1 is the original *explinandum,* and E_2, E_3, etc. are the new facts of geology, geographical distribution, morphology, embryology, etc.

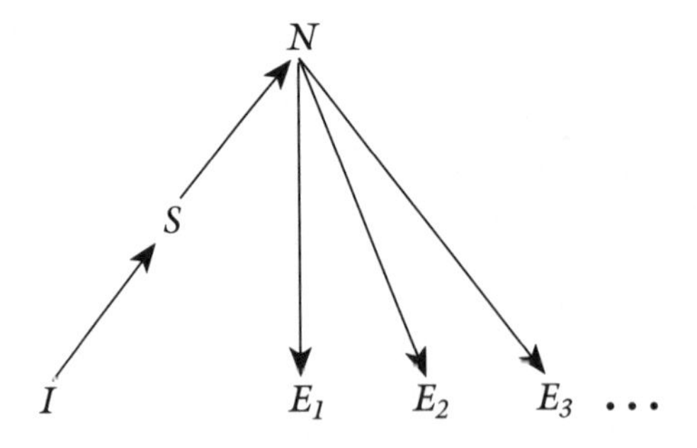

Nonetheless, analogy of hypotheses and consilience of inductions are by no means the only tools Darwin has at his disposal for making the concept of natural selection plausible as a *Vera Causa.* He uses several other arguments that are basically analogical, though of a different kind.

Arguments by double hierarchy. Immediately after introducing the concept of natural selection in chapter 3, Darwin reinforces *N* with the following argument:

(2.1)

> We have seen that man by selection can certainly produce great results, and can adapt organic beings to his own uses, through the accumulation of slight but useful variations, given to him by the hand of Nature. But Natural Selection, as we shall hereafter see, is a power incessantly ready for action, and is as immeasurably superior to man's feeble efforts, as the world of Nature is to those of Art.[59]

An argument along the same lines can be found in chapter 4 where Darwin writes:

(2.2)

> As man can produce and certainly has produced a great result by his methodical and unconscious means of selection, what may not nature effect? Man can act only on external and visible characters: nature cares nothing for appearances, except in so far as they may be useful to any being. She can act on every internal organ, on every shade of constitutional difference, on the whole machinery of life. . . . Can we wonder, then, that nature's productions should be far "truer" in character than man's productions; that they should be infinitely better adapted to the most complex conditions of life, and should plainly bear the stamp of far higher workmanship?[60]

Both these arguments follow the same pattern: if *B* (breeder) can produce *D* (domestic variations), why should *N* (Nature), far superior and more powerful than *B,* not be able to produce *E* (formation of new species)? The form these arguments take is slightly different from the others listed above in that they hinge on hierarchy rather than on analogy. Similarly, their aim is slightly different: while (1.1) and (1.2) try to make *N plausible,* or "probable" as explicitly stated in (1.2), (2.1) and (2.2) try rather to reinforce it, thus making it *more probable.* It is important to note, however, that this reinforcement has its price. As can be seen from a correction in later editions of the *Origin,* the more Darwin relies on analogy and hierarchy the more he runs the risk of personifying Nature ("Nature, if I *may be allowed* to personify the natural preservation or survival of the fittest"), thereby exposing himself to objections of a philosophical or theological order. Thus the very argument that is supposed to bolster the value of *N* in his interlocutors' eyes, ends up by weakening it.

Darwin has recourse to the same argument by hierarchy in still another context. This time he aims not at reinforcing *N* as such, but at explaining a given phenomenon in terms of *N*. The point in question is: how can Natural Selection have brought about such a complex instrument as the eye from a simple optic nerve? Darwin has no difficulty in admitting that this seems "absurd in the highest degree,"[61] but he firmly believes that the difficulty "though insuperable by our imagination" can be overcome. He begins by comparing the eye to a telescope; he then evokes innumerable variations which bring about "almost infinite" alterations "each improvement" of which "natural selection will pick out with unerring skill." Finally, he concludes:

(2.3)

> Let this process go on for millions on millions of years; and during each year on millions of individuals of many kinds; and may we not believe that a living optical instrument might thus be formed as superior as to one of glass, as the works of the Creator are to those of man?[62]

Here too the pattern is the same: if an effect (telescope) descends from a cause (craftsman), an analogous and superior effect (eye) can descend from an analogous and superior cause (Creator). And here too Darwin exposes himself to the criticism of transforming blind nature into an intentional Creator.

I shall now consider some of the arguments in set B whose goal, as already mentioned, is to defend the theory. At the beginning of chapter 6, Darwin lists four difficulties and objections: the lack of transitional forms; organs of trifling importance or of high complexity; instincts; sterile species. These facts belong to *explanandum* E and seem to disprove *N*. Darwin believes, however, that they are not so devastating and tries to show that they too can be explained by *N*. How does he do this? Relying on the hypothetico-deductive method, he would have to demonstrate that $N \rightarrow E$ or that $(N \wedge A) \rightarrow E$, where *A* is an auxiliary hypothesis. Darwin does on occasion follow this path. When, for example, he considers the question of transitional forms, which he himself describes as "the most obvious and gravest objection,"[63] he resorts to the hypothesis that "the chance of discovering in a formation in any one country all the early stages of transition between any two forms, is small, for the successive changes are supposed to have been local or confined to some one spot."[64] But this line of argument does not prove to be always fruitful. For lack of stronger deductive or hypothetico-deductive arguments, he is, then, obliged to use the rhetorical ones. Let us examine just a few.

(3) *Pragmatic arguments.* The question of the formation of the eye is so problematic that an argument by hierarchy alone is not enough. How can a simple nerve become sensitive to light and then develop into such a complex organ? Darwin first answers by setting down several facts supporting the existence of such a development; then he concludes:

(3.1)

> With these facts, here far too briefly and imperfectly given, which show that there is much graduated diversity in the eyes of living crustaceans, and bearing in mind how small the number of living animals is in proportion to those which have become extinct, I can see no great difficulty (not more than in the case of many other structures) in believing that natural selection has converted the simple apparatus of an optic nerve merely coated with pigment and invested by transparent membrane, into an optical instrument, as perfect as is possessed by any member of the great Articulate class.
>
> He who will go thus far, if he find on finishing this treatise that large bodies of facts, otherwise inexplicable, can be explained by the theory of descent, ought not to hesitate to go further, and to admit that a structure even as perfect as the eye of an eagle might be formed by natural selection, although in this case he does not know any of the transitional grades. His reason ought to conquer his imagination.[65]

What Darwin is wondering here is, in point of fact, the very opposite. He himself appeals to imagination to overcome *two* obstacles: one stemming from reason, that is, the inability to deduce the complexity of an eye uniquely from the premises of the theory; the other stemming from observation, that is, the lack of transitional grades. The argument is a classic example of the saying *allez en avant, la foi vous viendra.* Darwin invites his skeptical interlocutors to pursue the inquiry in order to gain future benefits or at least not lose advantages already acquired. Actually, his is an invitation to faith, in the hope that faith will conquer all resistance in the imagination and eventually convince reason.

(4) *Absurd and ridiculous.* Consider now Darwin's answer to the objection about seemingly harmful organs, or organs that appear to benefit other species. The presence of these organs, he says, "would annihilate my theory."[66] One obvious instance is the rattlesnake, but Darwin retorts:

(4.1)

> It is admitted that the rattlesnake has a poison-fang for its own defence and for the destruction of its prey; but some authors suppose that at the same time this snake is furnished with a rattle for its own injury, namely, to warn its prey to escape. I would almost as soon believe that the cat curls the end of its tail when preparing to spring in order to warn the doomed mouse.[67]

Here, clearly, the objection is not confuted. It is simply turned round and discredited by the technique of ridicule through a familiar comparison. The aim of this argument is also to gain time and to convince the interlocutor to go ahead despite the difficulties.

In chapter 5, Darwin resorts to this technique again, this time regarding the reversion to long-lost characters such as, for example, stripes in a horse:

(4.2)

> He who believes that each equine species was independently created, will, I presume, assert that each species has been created with a tendency to vary, both under nature and under domestication, in this particular manner, so as often to become striped like other species of the genus; and that each has been created with a strong tendency, when crossed with species inhabiting different quarters of the world, to produce hybrids resembling in their stripes, not their own parents, but other species of the genus. To admit this view is, as it seems to me, to reject a real for an unreal, or at least for an unknown cause. It makes the works of God a mere mockery and deception; I would almost as soon believe with the old and ignorant cosmologists, that fossil shells had never lived, but had been created in stone so as to mock the shells living on the sea-shore.[68]

(5) *Arguments by division.* A similar case is the bee's sting. Darwin's response is the following:

(5.1)

> If we look at the sting of a bee, as having originally existed in a remote progenitor as a boring and serrated instrument, like that in so many members of the same great order, and which has been modified but not perfected for its present purpose, with the poison originally adapted to cause galls subsequently intensified, we can perhaps understand how it is true that the use of the sting should so often cause the insect's own death; for if on the whole the power of sting-

ing be useful to the community, it will fulfil the requirements of natural selection, though it may cause the death of some few members.[69]

A similar argument can be found concerning the objection that the working ant is sterile and cannot pass on to its offspring any modification in structure or instinct: "it may well be asked, how it is possible to reconcile this case with the theory of natural selection?"[70] Schematically, here we have: $(N \rightarrow E) \wedge \neg E$. In order to defend the core of his theory, Darwin could either modify N or add some auxiliary hypothesis to it, or act on E. Darwin does neither, pointing out rather that,

(5.2)

This difficulty, though appearing insuperable, is lessened, or, as I believe, disappears, when it is remembered that selection may be applied to the family, as well as to the individual, and may thus gain the desired end.[71]

Both (5.1) and (5.2) are arguments by division. Darwin suggests that a variation which is useful to a family or a community is also useful to each of its members. The objection is thus dispelled, but another difficulty crops up: both arguments shift the biological subject to which the theory of natural selection originally refers from an individual to a group. This strategy is skillful, but the price is high.

We now come to some other arguments in set B. Objections to Darwin's *Origin* can be divided into two groups. Certain authors cast doubt on Darwin's method, while others were suspicious of the contents of his theory. In later editions of the *Origin,* he tried to respond to both. Let us start by analyzing the doubts cast on his method.

Although Darwin made every effort in his arguments (1) and (2) to provide stringent proofs of his hypothesis of natural selection, it could only be considered to be inductively derived from these facts. Darwin admitted this clearly in a letter to Asa Grey of November 20, 1859.[72] For followers of the Baconian method, however, this was a considerable defect. Sedgwick wrote that "Darwin's theory is not *inductive,* — not based on a series of acknowledged facts pointing to a *general conclusion,* — not a proposition evolved out of the facts, logically, and of course including them."[73] Hopkins compared Lamarck's and Darwin's theories with the theory of gravitation and the undulatory theory of light, observing that supporters of the former are content "to say that it *may* be so, and thus to build up theories based on the bare possibilities," while those of the latter "*prove* by modes of investigation which cannot be wrong, that phenomena exactly such as are observed would *necessarily,* not by the same vague possibility, result from the causes hypothetically assigned, thus demonstrating those causes to be the true causes."[74] Mill objected that it was not fair to accuse Darwin of having violated the rules of induction, but only because "Mr Darwin has never pretended that his doctrine was proved."[75]

Darwin was thus faced with two possibilities. He could either accept the accusation and admit that the hypothesis of natural selection had not been genuinely proved according to current standards, or he could reject the accusation and get involved in a debate over method. At the beginning it looked as though he was opting for the first possibility.[76] Later, however, he accepted the challenge, using the following arguments.

(6) *Arguments by retort.* In a letter to Lyell dated June 1, 1860, Darwin answered Hopkins's criticism in these terms:

(6.1)

> On his standards of proof, *natural* science would never progress, for without the making of theories I am convinced there would be no observation.[77]

In a later edition of the *Origin,* he added the following passage:

(6.2)

> In the literal sense of the word, no doubt, natural selection is a false term; but who ever objected to chemists speaking of the elective affinities of the various elements?—and yet an acid cannot strictly be said to elect the base with which it in preference combines. It has been said that I speak of natural selection as an active power of Deity; but who objects to an author speaking of the attraction of gravity as ruling the movements of the planets?[78]

In still another edition, he added the following remark to the "Recapitulation and Conclusion":

(6.3)

> It has recently been objected that this is an unsafe method of arguing; but it is a method used in judging of the common events of life, and has often been used by the greatest natural philosophers. The undulatory theory of light has thus been arrived at; and the belief in the revolution of the earth on its own axis was until lately supported by hardly any direct evidence. It is no valid objection that science as yet throws no light on the far higher problem of the essence or origin of life. Who can explain what is the essence of the attraction of gravity? No one objects to following out the results consequent on this unknown element of attraction; notwithstanding that

> Leibniz formerly accused Newton of introducing "occult qualities and miracles into philosophy."[79]

These are arguments by retort. Hopkins had once said, "he who appeals to Caesar must be judged by Caesar's laws."[80] Darwin's answer is that Caesar's laws favor him, not his opponent. He is not only asking for equal treatment (if Newton is allowed something, why can't I be?), but asserting that if this equality is not conceded, then his critics have to reject the conquests of science. In this way, Darwin retorts that the method on which his critics cast aspersions is the same method they themselves use and accept in practice. Their objection is thus autophagic, it refutes itself.

If we go on to examine the group of objections against the content of the theory, we find still other rhetorical arguments. Let us consider a few.

(7) *Arguments ad ignorantiam.* One of the most serious objections against the theory of natural selection was that of time. The geological lapse of time calculated by Darwin in chapter 9 of the *Origin* was much longer than the time calculated by Lord Kelvin, and the physical time was too short for the period of evolution Darwin needed. The objection was practically insuperable, but Darwin thought he could deal with it as follows:

(7.1)

> With respect to the lapse of time not having been sufficient since our planet was consolidated for the assumed amount of organic change, and this objection, as urged by Sir William Thomson is probably one of the gravest as yet advanced, I can only say, firstly, that we do not know at what rate species change as measured by years, and secondly, that many philosophers are not as yet willing to admit that we know enough of the constitution of the universe and of the interior of our globe to speculate with safety on its past duration.[81]

Although this might be considered a pragmatic argument, if we take it as an invitation to go ahead in the hope of benefits in the future, it seems more proper to consider it an argument based on ignorance. Darwin first claims that he who advances an objection must definitely prove it; he then observes that Lord Kelvin did not do so, and implicitly considers this a point in favor of his own theory. In short, his argument states: if you cannot prove your objections conclusively, you have to accept my claim.

(8) *Arguments ad hominem.* Most of chapter 7 of the final edition of the *Origin* is devoted to Mivart's objections. Darwin reports correctly that, in Mivart's view, species often manifest themselves "with suddenness and by modifications appearing at once," and that they often change through

an "internal force or tendency." To this he provides two answers. Here is the first:

(8.1)

My reasons for doubting whether natural species have changed as abruptly as have occasionally domestic races, and for entirely disbelieving that they have changed in the wonderful manner indicated by Mr. Mivart, are as follows. According to our experience, abrupt and strongly marked variations occur in our domestic productions, singly and at rather long intervals of time. If such occurred under nature, they would be liable, as formerly explained, to be lost by accidental causes of destruction and by subsequent intercrossing; and so it is known to be under domestication, unless abrupt variations of this kind are specially preserved and separated by the care of man. Hence in order that a new species should suddenly appear in the manner supposed by Mr. Mivart, it is almost necessary to believe, in opposition to all analogy, that several wonderfully changed individuals appeared simultaneously within the same district.[82]

The second answer goes like this:

(8.2)

He who believes that some ancient form was transformed suddenly through an internal force or tendency into, for instance, one furnished with wings, will be almost compelled to assume, in opposition to all analogy, that many individuals varied simultaneously. It cannot be denied that such abrupt and great changes of structures are widely different from those which most species apparently have undergone. He will be further compelled to believe that many structures beautifully adapted to all the other parts of the same creature and to the surrounding conditions, have been suddenly produced; and to such wonderful co-adaptation, he will not be able to assign a shadow of an explanation. He will be forced to admit that these great and sudden transformations have left no trace of their action on the embryo. To admit all this is, as it seems to me, to enter into the realms of the miracle, and to leave those of Science.[83]

Both these arguments can be classified in different ways: for example as arguments *ad ignorantiam* (taking them in the sense that either Mivart proves his own hypothesis or the theory of natural selection must be admitted), or as arguments by analogy (taking them in the sense that species cannot change suddenly because similar changes in the domestic state are soon lost). It seems to be better, however, to consider them *ad hominem* arguments, although of a special kind. Darwin, in fact, is not asking Mivart to prove his hypothesis, which is as Darwin himself admits in (8.1) ("natural species have changed abruptly as have occasionally domestic races"), partially supported by at least one analogy. He insists on miracles: to admit that many individuals could undergo variations suddenly and simultaneously through an internal force amounts, for Darwin, to entering into "the realms of miracle." This is why the argument can be considered *ad hominem*. Its purpose is to combat the adversary with his own arms, his own theses: Mivart claims to be a scientist and therefore cannot rest content with miraculous explanations. And yet, the argument is *ad hominem* in a sense that weakens its efficacy: Darwin is not, in fact, countering a statement of principle uttered by Mivart ("I am a scientist") with a thesis actually stated by him ("I admit miracles"); he is instead placing a statement of principle that can correctly be attributed to Mivart against a thesis he is only *supposed* to maintain. In his reconstruction of Mivart's position, Darwin is clearly forcing his point: "he who believes . . . will be *almost compelled* to assume . . . He will be further *compelled* to believe . . . He will be *forced* to admit. . . ."

(9) *Arguments of the possible.* Some of Jenkin's objections were particularly penetrating. In his attempt to deal with them, Darwin was often forced to make concessions; he always did so, however, by defending himself and trying to demonstrate that his theory was better. Lacking stronger proof, Darwin had no better instrument at hand than rhetoric for doing this. Let us consider some of his arguments in detail.

One of Jenkin's objections is particularly forbidding because it challenges Darwin on his own territory, that is, the analogy between domestic and natural selection. According to Jenkin, "although many domestic animals and plants are highly variable, there appears to be a limit to their variation in any one direction."[84] One can obtain, that is, new *varieties* but not new *species*. This was Darwin's rejoinder:

(9.1)

> Some authors have maintained that the amount of variation in our domestic productions is soon reached, and can never afterwards be exceeded. It would be somewhat rash to assert that the limit has been attained in any one case; for almost all our animals and plants have greatly improved in many ways within a recent period; and this implies variation. It would be equally rash to assert that characters now increased to their utmost limit could not, after remaining fixed for many centuries, again vary under new conditions of life.[85]

Another of Jenkin's criticisms was levelled at Darwin's use of the geological record to overcome the objection about the lack of transitional forms. As we have seen, in the first edition of the *Origin* he had already adduced the "extreme imperfection of the geological record,"[86] and in a later edition he added:

(9.2)

It is a more important consideration, leading to the same result, as lately insisted on by Dr. Falconer, namely, that the period during which species underwent modification, though long as measured by years, was probably short in comparison with that during which it remained without undergoing any change.[87]

Another of Jenkin's objections had its roots in Darwin's theory of blending inheritance. Given that theory, we cannot explain how a new useful character can become stable, because it would blend with others in later generations and thus soon disappear. Darwin admitted that "the justice of these remarks cannot, I think, be disputed";[88] but he did formulate an answer. On one occasion he wrote:

(9.3)

There must be some efficient cause for each slight individual difference, as well as far more strongly marked variations which occasionally arise; and if the unknown cause were to act persistently, it is almost certain that all the individuals of the species would be similarly modified.[89]

On another:

(9.4)

It should not, however, be overlooked that certain rather strongly marked variations, which no one would rank as mere individual differences, frequently recur owing to a similar organization being similarly acted on—of which fact numerous instances could be given with our domestic production. In such cases, if the varying individual did not actually transmit to its offspring its newly acquired character, it would undoubtedly transmit to them, as long as the existing conditions remained the same, a still stronger tendency to vary in the same manner. There can also be little doubt that the tendency to vary in the same manner has often been so strong that all individuals of the same species have been similarly modified without the aid of any form of selection.[90]

Three considerations must be taken into account here.

The first is that these are *arguments about what is possible.* In his attempt to deal with serious objections, Darwin introduces ad hoc modifications to his theory to show that, over and above all the evident difficulties, the theory is still able to offer *possible* or *conceivable* explanations.[91] Since these arguments do not claim to prove the truth of the theory, they do not commit their author to much. For them to be acceptable, however, they must satisfy at least two conditions: (i) that the possibility does not become an epistemically stronger modality unless additional reasons (such as new empirical evidence) are produced; (ii) that the possibility is consistent with other parts of the theory, in particular with its core.

The second consideration is that Darwin's arguments do not satisfy the first of these two conditions. Rhetoric seems to have led him astray in this case. If we consider his arguments in the order in which they are presented, it is clear that they move towards an epistemically stronger modality. An explanation that is at first "not impossible" in (9.1) becomes "probable" in (9.2), "almost certain" in (9.3) and "undoubtedly [certain]" in (9.4). Sometimes this crescendo takes place within the same passage, such as in the following:

(9.5)

> In many cases we are far too ignorant to be enabled to assert that a part or organ is so unimportant for the welfare of a species, that modifications in its structure could not have been slowly accumulated by means of natural selection. In many other cases, modifications are *probably* the direct result of the laws of variation or of growth, independently of any good having thus been gained. But even such structures have often, as we may feel *assured,* been subsequently taken advantage of, and still further modified, for the good of the species under new conditions of life.[92]

Although Darwin has several reasons for championing these possibilities, this gradual shift has no justification, and, worse still, it introduces new elements of difficulty in his theory.

The third consideration regards these very difficulties. Darwin's arguments for the possible do not even satisfy the second of the aforementioned conditions. His (9.1) and (9.2) conflict with the idea of gradualism, because they both suggest that evolution takes place over long periods of stability alternating with phases of variation; (9.3) and (9.4) conflict with the idea of spontaneity because they both admit that variations are brought about by modifications in the environment, even "without the aid of any form of selection"; (9.5) conflicts with the idea that all variations are beneficial to the individual in which they present themselves, because it admits there can be variations "independently of any good having thus been gained." In some cases, the contradiction is even more glaring: just one line after (9.3), Darwin writes that he had previously "under-rated, as it now seems probable, the frequency and importance of modifications due to spontaneous variability."

In his review of the *Origin,* Jenkin had written:

> The chief argument used to establish the theory rests on conjecture. Beasts may have varied; variation may have accumulated; they may have been permanent; . . . We are asked to believe all these maybe's happening on an enormous scale, in order that we may believe the final Darwinian "maybe" as to the origin of species. The general form of his argument is as follows:—All these things may have been, therefore my theory is possible, and since my theory is a possible one, all those hypotheses which it requires are rendered probable. There is little direct evidence that any of these maybe's actually *have been.*[93]

Must we then conclude, as several modern interpreters have done, that Darwin's "one long argument" is "a little loose,"[94] or that Darwin's "essential method was neither observing nor the more prosaic mode of scientific reasoning, but a peculiarly imaginative, inventive mode of argument,"[95] or that "Darwin's argumentative reasoning leaves much to be desired"?[96] The answer is, yet again, both yes and no. *Yes,* because judged according to the abstract model of deductive and inductive inferences, Darwin's "one long argument" does in fact take many weak and unjustifiable steps. *No,* because hypothetico-deductive and inductive inferences are not the only argumentative arrows in a scientist's quiver.

Not unlike Galileo, who proceeded not only on the basis of "sensory experiences and necessary demonstrations," Darwin did *not* rely uniquely on "true Baconian principles" or on the principles of the hypothetico-deductive method. Darwin too preached one thing and practiced another. If he had not, his *Origins* would never have been written, or it would have fallen prey to the first objections raised. At least all those who believe that the theory of natural selection is the pride of modern biology should be grateful to him for having violated the rules of method and taken up the arms of rhetoric.

3.3 Modern Cosmologists' "Observations and Calculations"

Modern cosmological texts, almost without exception, tell the same story. In short, it goes like this. Once upon a time there were two rival theories, the steady-state and the big-bang theory. For a while the two were at odds with each other, until one fine day two researchers from the Bell Laboratories, scrutinizing the skies for a completely different reason, bumped into a fact—background cosmic radiation—that offered crucial evidence against the first theory and in favor of the second. From that day on, cosmologists have lived happily with the winner of the two theories.

Like all good stories, this one is highly edifying but not very reliable, not so much because it does not describe the change of heart of cosmologists (things took place more or less in this way), but because it fails to describe *how* that change took place. With the aim of showing the role of rhetoric in science, I shall examine the question in more detail. Fred Hoyle has put it very well: "What happens when the balance shifts so that one theory comes more and more into favor while its rivals gently subside? What has happened to change the situation? If there was a good argument before to support a now-discarded theory, why isn't the argument still good?"[97]

Scientists typically answer this question exactly as Galileo did in the past; they maintain there are such things as "observations and calculations" that oblige whoever has made them to shift consensus from one theory to another. The following passage is revealing: "It is a tribute to the essential objectivity of modern astrophysics that this consensus [over the standard model] has been brought about not by shifts in philosophical preference or by the influence of astrophysical mandarins, but by the pressure of empirical data. . . . If some day the standard model is replaced by a better theory, it will probably be because of observations or calculations that drew their motivation from the standard model."[98]

Another cosmologist expresses the same view with some complacency for the difference between science and opinion:

> It is no longer necessary for scientists to have beliefs about when the creation of the universe occurred or what form it took (though they do have beliefs), it is now a question of using scientific instruments to see what the universe is like and how it has evolved. Such great philosophical issues are not debated as acts of faith, but as questions of evidence and theory, in the same way as other scientific disciplines. It is true that much of the present understanding of cosmological matters is rudimentary and tentative; certainly great upheavals in the currently accepted picture of the cosmos are very likely in the future. Nevertheless, it is important to appreciate that we are dealing here with science, and scientific values, so that while personal religious or philosophical preferences may make a great contribution to a particu-

lar individual's conception of the universe, the topics to be discussed in this book deal solely with concrete observational data and the controversies which rage around their theoretical interpretation.[99]

But is it really true that the influence of "astrophysical mandarins" is the only alternative to "observations and calculations," as Weinberg claims? And is it really true that "evidence and theory" are enough to distinguish personal opinion from scientific cosmology, as Davies upholds?[100] In the two previous sections we have seen that when the "pressure of empirical data" is not enough to shift consensus from one theory to another, the shift does not necessarily depend on power and subjective factors. In spite of philosophers and historians, modern cosmology provides us with further confirmation that the Cartesian dilemma is untenable.

In order to examine our two rival theories a little closer, let us take a picture of them at a time when significant empirical data were still missing—say at the end of the 1950s. Both the steady-state theory (*Ts*) and the big-bang theory (*Tb*) explain certain facts and certain astrophysical and astronomical laws, while presenting certain disadvantages. *Ts*, in particular, violates the principle of energy conservation because it calls for the continual creation of matter out of nothing; it lacks field equations (at least in Bondi and Gold's original version) and does not explain why the universe is expanding; it does not specify clearly the limits within which the universe is considered to be stable, that is, if these limits stretch out towards every conceivable confine or whether they stay within the nearby galaxies; it does not, finally, provide a satisfactory explanation of such phenomena as the overabundance of helium. *Tb*, on the other hand, does not explain the seemingly different ages of the galaxies and seems to be contradicted by the synthesis of heavy elements in the stars. Moreover, the two theories have different philosophical, metaphysical, and religious commitments, for example, as far as the creation of the universe is concerned: *Ts* thinks this problem can be reduced to a scientific matter, while *Tb* leaves the whole issue unsolved.

In a situation of this kind, "observations and calculations" clearly do not provide a sufficiently strong basis for deciding which theory is preferable or with which theory it is rational to work, since empirical advantages and disadvantages are more or less equally divided between the two. Some other basis must be found; most of all, some reasons must be advanced in order to argue that such a basis is sound. But how? Typically, one party would appeal to beliefs, values, or theories generally accepted by the community, and on such a basis, would try to persuade the other party that one theory is more promising because it is more compatible with certain known results, or because it explains certain facts considered particularly significant in the field, or because it satisfies certain basic assumptions, and so on and so on, until, to repeat Hoyle's words, "the balance shifts so that one theory comes more and more into favor while its rivals gently subside."

A heated debate held in 1959 between supporters of *Tb* and *Ts* proves to be particularly helpful in highlighting the nature of these arguments. In referring to it, I shall, as always, limit myself to a few examples.

(1) *Arguments ad ignorantiam.* W. Bonnor, who defends relativistic theories, claims that the universe has an unlimited past and future:

(1.1)

> This may seem in some ways as puzzling as if its history were finite. From the scientific aspect, however, this point is really one of methodology. Science should never voluntarily adopt hypotheses which restrict its scope. Sometimes restrictions are obligatory, as for example in the case of the uncertainty principle, which restricts the accuracy of certain physical measurements, but unless it is shown that such limitations apply to cosmology we should, I think, assume that our knowledge of the universe can stretch indefinitely into the past and into the future.[101]

In this argument, note the appeal to a methodological rule, namely, science should never adopt hypotheses that restrict its scope. It is interesting that this rule is explicitly introduced by an unless-clause with unspecified limits. This appeal, however, is placed in a rhetorical context; Bonnor claims that, unless the rival theory is proved, his own thesis should be accepted as true, or at least worth working on.

(2) *Arguments from authority.* Bonnor defends his relativistic theory of the universe in many different ways. One argument is that it is "in satisfactory agreement with present observations";[102] another is:

(2.1)

> Finally, let me stress that this theory is not constructed *ad hoc* to deal with cosmology. It is based on general relativity, which is known to be a satisfactory theory on a terrestrial scale and for the solar system. This gives one, I think, an added confidence in it.[103]

Here there is an (explicit) reference to a methodological rule ("do not introduce *ad hoc* hypotheses") and another (implicit) reference to a rule of inductive logic ("only generalize on a representative basis"). On the whole, however, Bonnor argues from authority. He claims that the steady-state theory is worth following because it is based on the general relativity theory, which is universally accepted. Confidence in the father is transferred, as it were, to the sons.

(3) *Arguments from a dilemma.* When defending the steady-state theory, Bondi introduces the perfect cosmological principle from which he claims the theory is derived:

(3.1)

> Of course, it may be necessary to consider the very difficult problems of the variation of physics in a varying universe; but before we enter the enormous complication of this question, we first try to see whether our universe might not happen to be one that is the same everywhere and at all times when viewed on a sufficiently large scale. In examining this possibility, we by no means claim that this must be the case; but we do say that this is so straightforward a possibility that it should be disproved before we begin to consider more complicated situations.[104]

This is a complex argument, for it contains both a dilemma and an appeal to ignorance. Bondi presents his interlocutors with two horns, and suggests that the first is possible whereas the second brings insurmountable problems that cannot be dealt with successfully. He then insinuates that, if the first horn is not disproved, there is good reason to pursue it. Bondi presents a further dilemma in the argument that follows:

(3.2)

> Either the laws of physics, as we have seen them here and now, apply everywhere and at all times, because the universe has been the same at all times and is the same everywhere, broadly speaking, or cosmology is a very much more difficult subject than I would like to tackle.[105]

Similarly:

(3.3)

> If the perfect cosmological principle turned out to be false, then cosmology would be a far more difficult subject than you seem to imagine. One would have to contemplate changes in local physics conditioned by changes in the universe and reacting back on it in an exceedingly complicated way.[106]

In all three cases, Bondi describes the setting as if there were only two exits, and then tries to convince his interlocutors that one of the apparent ways out is in fact a dead end while the other is almost obligatory since the path that leads to it (the perfect cosmological principle) is an a priori condition of cosmology itself. However, not only is the dilemma left unproven, but by taking the first horn Bondi disregards the fact that his methodological rule (not to introduce over-complicated hypotheses) has a consequence — in this case the continual creation of matter — that raises empirical and conceptual difficulties.

(4) *Pragmatic arguments.* Bonnor takes up these difficulties:

(4.1)

> According to the steady-state theory, matter is being continually created out of nothing in an empty space. Now we know from special relativity that matter is a form of energy, and so it follows that energy is being created out of nothing. But this infringes what we call the principle of conservation of energy, which has been confirmed by measurement to a high degree of accuracy and so is the principle that physicists in general will not abandon lightly. Bondi's view is that we have no evidence to suggest that a very slight rate of creation required by the steady-state theory does infringe the principle of conservation of energy within the limits of experimental accuracy, but I should have thought that, on the grounds of simplicity, it is much better to maintain that energy is accurately and exactly conserved. I think that we must demand a big dividend in return to justify our giving up this fundamental principle.[107]

This pragmatic argument tries to show that the steady-state theory causes serious upheaval in existing knowledge. Bonnor is arguing from the principle that hypotheses that conflict with evidence or well-grounded empirical laws should not be introduced. This rule is, however, debatable: first, one could claim that any revolutionary theory causes upheaval in significant segments of existing knowledge; second, it could be said that in this specific case the rule has not been violated in any significant way. In his reply, Bondi asserts just this, with the result that what for Bonnor represents a *disadvantage,* for Bondi becomes an *advantage.* His rejoinder is delivered to yet another pragmatic argument:

(4.2)

> Dr. Bonnor has argued that this process of continual creation violates the principle of conservation of energy which has withstood all the revolutions in physics in the last sixty years and which most physicists would be prepared to give up only if the most compelling reasons were presented; but this seems to me to be unsound. . . . Now, in fact, the mean density in the universe is so low, and the time scale of the universe is so large, by comparison with terrestrial circumstances, that the process of continual creation required by the steady-

state theory predicts the creation of only one hydrogen atom in a space the size of an ordinary living-room once every few million years. It is quite clear that this process, therefore, is in no way in conflict with the experiments on which the principle of the conservation of matter and energy is based. It is only in conflict with what was thought to be the simplest formulation of these experimental results, namely that matter and energy were precisely conserved. The steady-state theory has shown, however, that such simplicity can be gained in cosmology by the alternative formulation of a small amount of continual creation, with conservation beyond that.[108]

(5) *Arguments by definition*. While Bondi denies that the steady-state theory conflicts in any experimentally relevant sense with the principle of the conservation of energy, Lyttleton admits the conflict is genuine. He does not, however, claim that this principle is an untouchable dogma:

(5.1)

Of course, it is true that the principle of the conservation of energy has survived for a long time, but I think that Eddington put his finger on it when he pointed out that the reason why it has survived for so long is simply because in physics energy has come to be defined as that which is conserved. And what has happened is that from time to time new things have been introduced as energy to save this principle.[109]

This argument tends to suggest that a line of inquiry can be pursued even if it violates an accepted law, provided this law is not "frozen" into a definition. The argument implicitly refers to that (Popperian) methodological principle according to which scientific inquiry should never be defended from revision by using conventionalistic stratagems. Note, however, that Lyttleton does not reject definitions altogether; the continuation of his argument is based on just a definition:

(5.2)

Now, I cannot see that laws of physics that change with time are really laws of physics at all. This is perhaps an act of scientific faith on my part, but I think that we must always formulate scientific laws in such a way that time itself does not enter into them explicitly.[110]

Here, a methodological convention is used to reject a line of inquiry. The tactic is skillful because it creates from the outset a philosophical obstacle for the interlocutor, who is then forced to remove it before he can carry on with his thesis.

(6) *Argument of reciprocity.* Lyttleton stresses his aversion to seeing the principle of conservation turned into a definition:

(6.1)

In postulating a rate of creation that is smaller than the most refined measures of the law of conservation, no conflict with empirical evidence has been introduced at all. On the contrary, I would maintain that this is in accord with one of the typical ways in which science advances. You may remember that at one time in chemistry all the atomic weights were thought to be exact integers, but a great advance occurred when it was pointed out that it was not precisely true.[111]

This is an argument of reciprocity. It maintains that if a certain line of action is admitted in chemistry then it should also be accepted in cosmology. The argument also contains an implicit retort (if such and such a line of action is condemned in cosmology, then chemistry too is in error), an appeal to authority (scientific tradition), and a pragmatic suggestion (if chemistry has benefited from a certain line of action, then so will cosmology). Obviously Lyttleton's argument could be challenged in the manner of a judge's appeal to precedents, that is, either by questioning the symmetry of the situations, or by recalling past cases that received different treatment.

And this is precisely what Bonnor does. To Bondi's objection that the theory of general relativity should not be hastily extrapolated to conditions different from those it is based on, Bonnor rejoins:

(6.2)

I certainly believe that the field equations of general relativity are valid for all states of the universe. What we do in relativistic cosmology is to take laws which have been established for local gravitation of the universe as a whole. This has led to no conflict with observation, and is in any case well-established scientific practice.[112]

Here a scientist is in the same situation as a judge: in order to decide about a new case he looks up the precedents. Bonnor's view, however, is disputable and the cases that might comfort him are all too often in conflict. Scientific jurisprudence is no more univocal than legal jurisprudence.

What conclusions, then, can be drawn from this debate regarding Hoyle's question with which we started?

The first conclusion is that a shift in consensus in favor of a theory does not depend on the "pressure of empirical data" alone or on rules formulated only in terms of empirical data. This is not because these rules are not used: actually the debate reveals that the participants rely more or less explicitly on methodological rules (more precisely, on maxims that function as methodological rules). They do so because these rules work (when they actually do) only after the problems raised by their interpretation and application have been solved, that is, after certain preliminary decisions have been made. For example, Popper's rule, "Choose the theory that takes more risks," explicitly advocated by Bondi, presupposes a preliminary agreement over which facts are riskier and which tests are more severe. Similarly, Laudan's principle, "Choose the theory that, all things being equal, presents fewer conceptual problems," implicitly advocated by Bonnor, presupposes a preliminary decision about which conceptual problems are more or less important. To return to our example, *if* we agree that *Ts* is in conflict with *O* (e.g., the nonobservation of new matter out of nothing), with *L* (e.g., the law of the conservation of energy), and with *T* (e.g., the theory of general relativity); if we also agree that *O* is really the case, that *L* is well-established, that *T* is widely accepted, that explanation is a desirable aim of science, that external coherence (agreement of a theory with the facts) and internal coherence (agreement of a theory with itself and with already accepted knowledge) are equally desirable; and, if we agree, finally, that *Tb* explains *O* and does not conflict with *L* or *T*—*then* the shift in consensus from *Ts* to *Tb* is compelling and natural. The list of preliminary decisions without which maxims or rules are of no use is a long and winding road, with many obstacles along the way.

This point can be further illustrated by considering one of Bonnor's comments:

> I should think that the steady-state theory is most likely to be disproved by the radio observations on variations with distance. If it should survive a series of reliable observations of this type, it would be taken very seriously.
>
> On the other hand, the relativistic theories would be difficult to disprove definitely by this means. The reason is that these theories are not represented by a unique model of the universe, and models exist in which the variation of density with distance, though not quite negligible, is none the less small. Thus, as Bondi had said, this might simply mean that the astronomers were not yet probing far enough into space to notice any significant variation.[113]

Faced with this Lakatosian situation, a hair-splitter would say that it is easy to dictate rules fixing the conditions beyond which a program degenerates and it becomes irrational to pursue it. But this is clearly putting the cart before the horse. The crucial point here is to understand how these conditions can be established, and with what arguments.

This brings us to our second conclusion regarding the problem of the shift in consensus from one theory to another. Hoyle's solution to this problem is complex. In his view "there are at least three reasons for these shifts of emphasis."[114] These occur when new data confirm one of the two rival theories; when old data that conflict with the theory are reinterpreted; and when the theory is shown to be more probable. Considering the difficulties of proving this third reason, Hoyle concludes that "some skill and a lot of luck are needed if you are to avoid throwing away rough diamonds with the rest of the rubble."[115]

This answer is satisfactory but insufficient—not because skill and luck do not play an important role, but because they are no better than rules, in that they work only if they are well argued. Our analysis shows that these arguments do exist and that they are typically rhetorical. What has been said of Galileo and Darwin also applies here. Cosmologists profess a method, but in practice they either violate it or make every effort to stretch it to their own convenience. In this they are not so different from good lawyers who know the law like the back of their hands, but spend their time creating quibbles for the benefit of their clients.

3.4. The Functions and Techniques of Scientific Rhetoric

At the end of the previous chapter, I expressed two goals for the present one: to document the fact that scientists *do* use rhetoric and to understand *why* they do so. The first goal can be considered achieved. We have seen that, although Galileo argued against "flowers of rhetoric" and maintained that there is room only for "sensory experiences" in science; that although Darwin claimed he worked only according to "true Baconian principles"; and that although modern cosmologists affirm that the shift of consensus from one theory to another depends only on the "pressure of empirical data," there is no longer any doubt that scientists use typically rhetorical arguments in addition to deductive and inductive ones.

We are left with the second goal. Speaking in the abstract, I have mentioned several reasons why rhetoric plays a role in science. On a more concrete level, the debates among scientists examined in the last three sections help us to see the main functions of rhetoric in scientific contexts.

Choosing a suitable methodological procedure. This problem comes about when a new theory is associated with a new method. The case is rare, but it occurred with Galileo and Darwin, as well as with Bondi, whose methodological approach was different from that of his critics. The task of innovators in these cases is very difficult. They must commit themselves on two fronts: they know that their own theory has no chance of passing if the new method does not get through; they also know that the main obstacle against the new method is precisely the theory to which it is associated. In some cases they will start off with the theory; in others they will offer a "discourse on method." In the latter case, they will often use the technique of *ad hominem* argument to reveal a contradiction between what their adversaries *say* and what they *do,* making every effort to show that what they do is not only the opposite of what they say but the application of the very method they are rejecting. Some of the examples we have given confirm the frequent use of this technique. Let us return to them briefly.

Galileo has to deal with Simplicio, an interlocutor who does not accept the method of "sensory experiences and necessary demonstrations," and he has to convince him that it is a good method for reaching truth about natural matters. As we have seen from his arguments (1.1) and (2.1), Galileo tries to show that such a method is the same as the one professed or applied by the authority Simplicio bows to (Aristotle). Likewise, Darwin has to defend himself against critics who accuse him of using an unorthodox method (the method of hypotheses). He too has to convince his adversaries that his own method is a good one; he too tries to show that it has been adopted successfully in the past by recognized authorities (Newton). His arguments (6.1) and (6.3) have precisely this aim. Bondi's situation is not so different. As his deductive method (proving the steady-state theory starting with the perfect cosmological principle) contradicts the generally accepted inductive method adopted in cosmology, he has to show that his own method is a good one. Although his arguments are not evident in the debate we have examined, they can easily be documented from his other writings.[116]

The technique of so-called "circumstantial" *ad hominem* argument is often resorted to in cases where the protagonists are very far apart. This technique, in fact, does not attempt to prove a thesis, nor disprove that of an adversary, nor attack his person. Its aim is to create a breach, shake people's confidence, weaken their resistance. The argument: "How can you criticize the method of 'sensory experiences' (or the hypothetico-deductive method, etc.) and then practice it?" is the same as the argument "How can you call yourself a vegetarian and eat chicken every day?" Once the breach has been opened, the interlocutor should be ready to consider the thesis proposed.[117] Other arguments can then be used.

Interpreting a methodological rule. The problem here lies in establishing the exact prescriptive content of a methodological rule. These rules are never written explicitly. They are usually tacit norms or maxims of behavior derived from or connected with the epistemic values of the tradition learned during scientific training and practiced in the profession. This is why they often give rise to controversy. Two researchers may well agree on what rule is proper to the case at hand and yet have different opinions about how to interpret it. Consider the rejection rule. In one of its many formulations it states that if a hypothesis H has at least one logical consequence O_l, such that it must be rejected whenever $O_l \neq O_e$, if O_e is the uncontroversial result of a well-conducted experiment. Although this formulation is much debated, it is sufficiently general for nearly all researchers to agree on—in word if not in deed. For example, in our cosmological debate, Bondi and Lyttleton share this rule with their interlocutors. Yet, when they are asked to apply it, and therefore, for coherence's sake, to reject the steady-state theory that violates it, they refuse, providing instead different interpretations of the same rule. As argument (4.2) shows, Bondi interprets the rule in the sense that a theory T should *not* be rejected if $O_l - O_e = \varepsilon$, where ε is such a negligibly small quantity that it cannot be experimentally ascertained (as in the case of the creation of new matter out of nothing). Analogously, as argument (5.1) shows, Lyttleton interprets the rule in the sense that T should *not* be rejected even if $O_l - O_e = n$, with n being as big as anyone wants, provided O_e is not a genuine empirical fact but an implicit or disguised definition (as Lyttleton takes the law of the conservation of energy).

The typical rhetorical techniques in these cases are pragmatic arguments that appeal to precedents. A researcher who finds that his theory meets with empirical difficulties will try to mitigate the severity of the rejection rule by adducing the advantages that can be reaped from suspending its application, or by trying to show that in other cases benefits have been obtained by casting doubt on some prima facie conflicting empirical evidence. He behaves like a lawyer who tries to discredit the prosecution's witness or convince the jury that in similar cases in the past such witnesses have been given little credibility.

Applying a rule to a concrete case. A rule can be accepted and its interpretation agreed upon, but doubts may still remain as to its pertinence to the case in question. These doubts are often difficult to resolve, especially when the objection is raised that the case is different from the one contemplated in the abstract by the rule. Whoever raises these objections is like a lawyer who argues that a certain deed committed by his client does not fall under such and such a law, or that the application of such and such a law to this case would be unfair. An example is provided by Bonnor's argument (1.1) in which he tries to convince his interlocutors that cosmology should be no exception to the rule of not introducing restrictive hypotheses. The technique of the argument *ad ignorantiam* is used here to create a dilemma: either it is proved that cosmology is a special case or the general rule applies. If the interlocutor accepts the dilemma, and he has no convincing proof in favor of the first horn, at least the second will have the chance of being considered.

Rhetoric has more than these three typical functions, however. The main ones can be classified into the following categories.

Justifying a starting point. In order to achieve a certain result, one must start off with certain premises. The more the premises are shared, the more likely the acceptance of the result. But how can the premises be justified? As we have seen, Aristotle observed that dialectical arguments serve this purpose very well. If a premise is dubious, the interlocutors will have to be convinced to admit it, at least for a while, in the hope that once the result is achieved they will then be happy to accept it. The reasoning is based on the saying, "labor pains are soon forgotten." In this case, again, a useful technique is to create a dilemma. The researcher will try to reduce the situation to two possible solutions and show that the horn of the dilemma containing the rival solution leads to insurmountable difficulties. Bondi uses this technique in favor of the perfect cosmological principle. His arguments (3.1)–(3.3) try to show that, if this principle were denied, cosmology would no longer be possible or it would be so difficult that there would be no chance of successfully pursuing it.

Attributing to a hypothesis a positive degree of plausibility or reinforcing it. A hypothesis with a low plausibility (initial probability) will not be taken seriously into consideration. Furthermore, as Bayes's theorem shows, if a hypothesis does not have a positive degree of plausibility it cannot have a positive degree of confirmation. It is therefore essential to be able to attribute a hypothesis with a positive degree of plausibility or to reinforce it. This attribution can be carried out in a number of different ways. One can, of course, try to show that a hypothesis follows deductively from an accepted premise, as Bondi does with the hypothesis of the steady-state from the perfect cosmological principle; or that it follows inductively or analogically from certain empirical premises, as Darwin does in arguments (1.1)–(1.2) with the hypothesis of natural selection from domestic variations. But many other rhetorical moves are possible, none of which can be said to be typical. A researcher will try to show that his own hypothesis satisfies certain accepted requirements or that it does not contradict others. Another will argue that it is of the same type as others already accepted or that the doubts it raises can be left aside. Yet another will appeal to general methodological or ontological assumptions, as in Galileo's arguments (7.1) and (7.2). But the list of moves is still long. In order to persuade others that his own theory is plausible, a scientist will adduce general or particular properties of nature, as in Galileo's (8.1)–(8.2) and Darwin's (2.1)–(2.3); or he will invoke the same treatment reserved for others, as in Darwin's (6.1) and in Bonnor's (6.2); or he will call on the authority of certain precedents considered analogous, as in Bonnor's (2.1); or he will invite his interlocutors to have faith and carry on, as in Darwin's (3.1).

Criticizing or discrediting rival hypotheses. The best way to criticize a scientific hypothesis is to show that it conflicts with generally accepted data. But this is not always as easy as it sounds. During a controversy, the proponent of a hypothesis is unlikely to admit the situation, and if he does, he will find arguments for reducing the negative impact of this conflict. He will claim, for example, that the data have been interpreted wrongly or that the difficulties can be overcome or that they are not so serious as to require the rejection of the hypothesis. The discussion will thus change in tone and the critic will alter his register, moving from empirical difficulties to other kinds of difficulties. For instance, he will try to show that the hypothesis is not as fertile or simple as its proponent makes out: this is the case of Bonnor's argument (4.1) against the steady-state theory. Or else he will use the technique of *ad hominem* argument, as do Galileo (4.1) and Darwin (8.1) and (8.2), trying to show that the hypothesis in question is contrary to others already explicitly or implicitly admitted by the proponent during the discussion. Or again, he will try to lead the hypothesis towards the absurd and ridiculous, as do Galileo (11.1)–(11.5) and Darwin (4.1) and (4.2). Finally, when he has no better arms with which to fight, he will argue *ad personam,* as does Galileo (5.1), in order to discredit his adversary, precisely like a lawyer who, during a cross-examination, begins by trying to confuse the defendant with facts or evidence against him, goes on to demonstrate the scant plausibility of what he is saying, and ends up by discrediting him altogether by commenting on his personal habits or inclinations.

Rejecting objections against a hypothesis. For a hypothesis to be successful in a discussion, it is not enough for the proponent to put his adversaries in a tight spot; he must also refute their criticism. A scientific dispute is a shooting match; it rarely ends up with a death, but more often with points on each side. The winner is the one who has fired the most shots or the one who has aimed best or the one who has received the least wounds. Since the exchange is reciprocal, the arms used to criticize a hypothesis can also be used to defend it. Attempts will be made to convince the interlocutor that the hypothesis is well confirmed, then to show that it does not meet with significant anomalies, and finally to weaken the possible conceptual difficulties deriving from its conflict with accepted theories or admitted worldviews. With this aim in mind, the usual techniques will be employed: from substantive considerations on the nature of the reality under scrutiny, as in Galileo (3.1) and Darwin (5.1) and (5.2); to analogies and comparisons, as in Galileo (6.1) and (6.2); to pragmatic arguments, as in Galileo (9.1) and Bondi (4.2); to arguments *ad ignorantiam,* as in Darwin (7.1). Sometimes, when the task is more difficult and the proponent of a hypothesis is put into a corner and is weighed down with the criticism, he will try to defend his proposal by admitting the existence of local difficulties, but he will nevertheless try to show that the hypothesis is at least possible, in the hope that during the discussion he will be able to transform this possibility into something epistemically stronger. As we have seen, this is the case in Darwin (9.1)–(9.5), where the transformation takes place surreptitiously without new reasons being recognized by the adversary.

The functions examined here by no means exhaust all the possible roles rhetoric plays in science, but they are sufficient to conclude that this role is not merely ornamental. It could be objected that, in science, rhetorical techniques are resorted to—as the cosmologist D. Sciama put it—"in the middle of the debate," when "we do not know the ultimate outcome and we must be guided by our own sense of the fitness of things,"[118] because "ultimately, the only test is the pragmatic one of whose ideas succeed the best." In a similar vein, another cosmologist, J. Narlikar, has written that "history of science shows that theories are not discarded on a majority vote but only when observations are unambiguously against them."[119] But what does "ultimately" mean? Does it mean when we have done *all* the tests? As Bondi once wrote, "we can never wait until we have all the facts at our disposal; that time never comes."[120] And what does it mean for a fact to be "unambiguously" against a theory? Does it mean when it is manifestly against it? This time, too, never comes. In science, we are always "in the middle of the debate." One can get out, of course, but, as with all debates, one can get out only by producing good reasons.

Since establishing these reasons is the task of scientific dialectics, it is now time we turn to it.

Scientific Dialectics

Moreover, as contributing to knowledge and to philosophic wisdom the power of discerning and holding in one view the results of either of two hypotheses is no mean instrument; for it then only remains to make a right choice of one of them.

Aristotle, *Topics*

4.1 Rhetoric and Change of Belief

The arguments examined in the previous chapter constitute a sample of what I have called "scientific rhetoric." Although they have different forms and follow different lines, all of them aim at convincing an audience, at obtaining consensus for a certain claim, be it the plausibility of a hypothesis, the intellectual and pragmatic advantages of a research program, the explanatory merits of a theory, or something else. We must now examine what kind of arguments these are, and with what canons they can be validated; that is, what kind of logic scientific dialectics is.

A widespread view in logic textbooks—call it "logical dualism"—maintains that arguments are either deductive or inductive; if they cannot be reduced to either a valid deductive or a correct inductive form, they are said to be fallacious.[1] It follows that deductive and inductive logic are the only tools for appraising arguments. In this view, the dialectics we are looking for could only be similar to what Kant described as "a sophistical art of giving to ignorance, and indeed to intentional sophistries, the appearance of truth, by the device of imitating the methodical thoroughness which logic prescribes, and using its 'topic' to conceal the emptiness if its pretensions."[2] Fortunately, logical dualism is attractive but it proves to be too narrow a view.

Take a dilemma like Bondi's first. Schematically, it can be put in this form:

(1)

> Either we accept the perfect cosmological principle, or, if physical laws change with time and space, cosmology is impossible to pursue.

This argument exhibits the valid deductive form "*p* or *q*, not-*q*; therefore *p*." Yet an argument with this form may be fallacious. This is the case if one can show there is a way of shifting between the two horns. It is not the case if the two horns are taken as incompatible and exhaustive. But considering the form of the arguments alone does not provide us with any information about which of these possibilities is the case, and therefore it is not enough to appraise it fully.

Take another argument. In the Second Day of the *Dialogue,* Galileo rejects the following argument put forward by Simplicio:

(2)

> Stones move downward because they possess gravity.

If "moving downward" and "gravity" are taken as synonymous, the argument is reducible to the form "if *p* then *p*," which is valid according to deductive logic. Yet if "gravity" is just a name for "moving downward," the argument cannot be used as a physical explanation because, as Galileo writes, "we do not really understand what principle or what force it is that moves stones downward,"[3] and therefore it is fallacious: it is a *petitio principii* whose conclusion merely restates the premise using different words. But in a situation in which the force of gravity was independently known, the argument would be perfectly acceptable. The same can be said of an argument like this:

(3)

> It rains because water falls from the sky.

Once the synonymy has been established, this too is trivially valid. However, the argument is fallacious if considered according to our standards of scientific explanation, while it is not fallacious if it performs a different function or if the standards are different, for example, if it is offered to, or put forward by, a child whose criteria for explaining phenomena are not the same as those dictating an adult's meteorology.[4] Here, too, considering the form of the argument alone is not sufficient to establish whether it is good or fallacious.

Take now an argument from authority of the following kind:

(4)

p because *X* says that *p*.

Although it is often treated as elliptical, and efforts are made to transform it into either a deductive or an inductive argument (by adding, respectively, "everything *X* says is true" and "most of the things *X* says are true"), it is not reducible, for it eludes the standards of deductive and inductive logic. As has been shown,[5] if another authority *Y* exists that maintains not -*p*, then we would have that *p* and not -*p* are both true and false (if the argument is transformed into a deductive one by adding the premise "everything *Y* says is true"); or that *p* and not -*p* are both probable and improbable (if the argument is transformed into an inductive one by adding the premise "most of the things *Y* says are true"). Yet the argument may not be fallacious if *X* is an authority competent in the domain to which *p* belongs, or if *p* is not asserted as true but simply suggested as worth working with. Once again, the form of the argument does not determine whether it is good or fallacious.

Finally, take a circumstantial ***ad hominem*** argument such as:

(5)

X maintains *p*, *X* practices *q*, *p* and *q* are incompatible; therefore not -*p*.

From the point of view of deductive logic an argument with this form is invalid. But it is not always fallacious. Remember Darwin's reply to W. Hopkins. If Hopkins accepts, considers, and works with theories referring to nonobservational entities such as the gravitation theory or the undulatory theory of light (*q*), then this is a good reason for rejecting his view (*p*) that natural selection is unacceptable since it contains similar entities. As in all circumstantial ***ad hominem*** arguments, the inconsistency between *p* and *q* is "pragmatic," but here it is so strong as to be almost theoretical, because while *p* affirms that theories in nonobservational terms are not to be accepted, the practice *q* amounts to claiming that certain theories in nonobservational terms are acceptable.

These examples show that valid arguments may be fallacious, while arguments that are invalid (or not reducible to valid ones) may not be fallacious.[6] Logical dualism therefore is to be abandoned. This comes as no surprise, because logical dualism is a typical by-product of those rationalistic as well as empiricist views according to which everything outside relations of ideas (formal deduction) and matters of fact (inductions from experience) is to be discarded as irrational.[7] If these views pass a sponge over rhetorical arguments and make them all trivially fallacious it is because they are not open to other forms of rational reasoning. To understand the nature of rhetorical arguments we have to widen our idea of rationality and be prepared to find a proper logic for them rather than reject them because they do not fit clearly with ready-made logic. What is essential is to compare them with other arguments as regards their purpose.

Take a deductive argument first. It aims at showing that a certain conclusion necessarily follows from a given set of premises or that a given proposition can be derived from a certain set of premises. Deductive logic fixes the conditions of the validity of this derivation. It does not consider the truth value of the premises or the conclusion of the argument; nor does it care about the credibility of the conclusion for a given person. Actually, deductive arguments have no interlocutor or they are taken as being addressed to a "universal audience," which amounts to the same thing. If a deductive argument is valid according to deductive logic, then it is valid even if its conclusion cannot be included in the body of our beliefs.

Consider now a typical inductive argument like "All observed *A* have been *B*, therefore all *A* are *B*." Its purpose is to infer a conclusion that goes beyond the evidence mentioned by the premise. Since standard (qualitative) inductive logic has been traditionally conceived as weak deductive logic, it aims at establishing whether the inference is correct, whether its conclusion can be asserted given the evidence. It is true that, unlike deductive logic, inductive logic is interested in the intrinsic credibility of the conclusion of the inference, but this credibility is made to depend on the logical link between the conclusion and the premise of the argument. Inductive arguments, too, have a "universal audience." Thus, like deductive logic, inductive logic is universal in scope. If an inductive argument is correct according to inductive logic, then it is correct, whether it is "reasonable" for an interlocutor to accept its conclusion or not.

Rhetorical arguments have a different purpose. They are addressed to specific interlocutors with specific systems of expressed or implicit beliefs, and they aim at changing such systems. Thus rhetorical arguments are not good or bad in themselves; they are good or bad according to the given situations in which they are put forward and the specific audiences to which they are offered.

For this reason rhetorical arguments also differ from probabilistic arguments and cannot be treated with the tools of the logic of plausible reasoning such as Polya's.[8] This logic contains rules such as the following (where $\rightarrow$ means "implies" and $|$ means "incompatible with"):

$$\text{(I)} \qquad \begin{array}{l} p \rightarrow q \\ q \text{ true verified} \\ \hline q \text{ more credible} \end{array}$$

(II) $p \rightarrow q_n$
q_n different from $q_1, \ldots, q_{n-1}$ already verified
q_n true

---

q_n much more credible

(III) $p \rightarrow q$
q hardly credible without p
q true

---

p very much more credible

(IV) $p \mid q$
q false

---

p more credible

Rules like these show that the purpose of probabilistic arguments is to change the degree of credibility of a proposition given the truth, falsity, or the degree of credibility of other related propositions. And the logic of plausible reasoning aims at establishing the formal relationship between these degrees, that is, at establishing *relative* increments of credibility (its "direction" and "magnitude" as Polya writes),[9] not *absolute* attributions of credibility or *absolute* reinforcement or weakening. In this sense probabilistic arguments are like deductive and inductive arguments. Likewise for the logic of plausible reasoning, as shown by its formalization in the probability calculus, in particular Bayes's theorem:[10] if a probabilistic argument is correct according to this logic, then it is correct and stringent for all those who accept its premises.

Summing up, we can say that deductive arguments aim at *deriving* a conclusion from certain premises, inductive arguments at *inferring* a belief, plausible arguments at *reinforcing* or *weakening* the degree of credibility of a belief, and rhetorical arguments drive at *changing* the system of beliefs of an interlocutor during a debate. We can then give the following definitions of rhetoric: in its narrow sense, *rhetoric is the set of rhetorical arguments,* that is, those arguments that cannot be appraised with the tools of formal logic; in its broad sense, *rhetoric is the set of all arguments that aim at inducing a change in belief in an audience during a debate.*

4.2 Dialectics, Audience, and Debate

We are now in a position to define dialectics. If we take rhetoric in its narrow sense, we can say: *dialectics is the logic of rhetorical arguments.* If rhetoric is taken in its broad sense, we can say: *dialectics is the logic of belief change in an audience.*[11] If we say that *dialectics is the logic of debate,* this definition encompasses both senses.

As we have seen, formal logics (both deductive and inductive) make no reference to debate because they examine *arguments in themselves* and aim at establishing whether or not they are valid or correct according to certain rules of derivation. If p implies q, and p, then q necessarily follows according to deductive logic's rule of *modus ponens.* If all observed A have been B, then "All A are B" is probable according to inductive logic's rule of generalization. And if p implies q, q is true, and p has a positive degree of credibility, then "p is more credible" derives from the fundamental rule of plausible reasoning.

With dialectics the situation is different. Since it aims at establishing whether arguments are good or bad in specific situations for specific audiences, it must deal not with arguments in themselves but with *arguments in a debate.* An argument may be valid or correct when taken out of context but bad when considered in a debate; conversely, it may be invalid and incorrect when taken out of context but good when considered in a debate. The fact is that, as part of a debate, an argument is submitted to certain constraints or rules governing the debate and establishing which moves are prohibited or permitted. Dialectics fix such rules.

Consider a few examples. A dilemma such as "p or q, not-p; therefore q" is bad if another alternative r, resulting from the debate, is not taken into consideration: in this case, the argument violates the rule never to leave questions or objections unanswered (in this case the question "why not r?"). An argument such as "p because p" is bad because if an arguer states p then he cannot use p as an explanation of p without violating the rule always to give grounds for answers (in this case the answer to "why p?"). Many rhetorical arguments commit so-called "informal fallacies" because they violate rules of debate such as these.[12] Taken in the narrow sense of the logic of rhetorical arguments, dialectics is also the *logic of fallacies.*

But there is more. Rhetorical arguments aim at inducing change of belief. This does not mean, however, that only those arguments that cannot be dealt with by formal logics are rhetorical. *All* arguments are rhetorical if they are used rhetorically. For example, if A believes q, B can change this belief if he proves that $q \rightarrow r$ and $\neg r$, by using deductive logic. If A rejects q (say an empirical hypothesis), B may try to convince him to change this belief if he proves that all the ascertained observational consequences O_1, . . . O_n of q are true, thus using inductive logic. If A intends to convince B to work on q, B may try to change A's mind by showing him that q is not very credible because it derives from a premise that is false or incompatible with already accepted knowledge, thus using probabilistic logic. These arguments are valid or correct, but not necessarily good. Valid or correct arguments are good provided they are considered *pertinent* to the specific question under discussion and provided the logic with which they draw their conclusions is agreed upon. For example, an inductive argument showing that all the applications of a certain mathematical theorem have been verified would be fallacious if used to prove that theorem. Since dialectics is concerned with change of belief, it must also be concerned with the right tools for changing belief, that is, with the correct use of those tools. Thus, taken in the broad sense of the logic of belief change, dialectics is also the *logic of (the rhetorical use of formal) logics* or the logic of argumentation.[13]

In order to ascertain the right logic for an argument, an analysis of its structure is not enough. Only the context can provide the necessary information. Out of their contexts, taken as linguistic strings of premises and conclusions bound with link locutions, arguments are not dissimilar to a set of lines randomly drawn on a piece of paper. Just as these lines may allow many different readings when no background is given, so arguments are multivalent when no context is provided. Take, for example, an argument with this form: $((p \rightarrow q) \wedge q) \rightarrow p$. Should we say it is deductive and invalid according to deductive logic, or that it is inductive and correct according to inductive logic?[14] Only the context provides an answer. If it is used to *prove* a proposition *p*, then the argument is deductive and deductive logic is pertinent to it. If it is used to *confirm* a hypothesis *p*, then it is inductive and falls within the legislation of inductive logic. Thus the very same argument with the very same form is potentially fallacious if it is used for one purpose and potentially good if used for another.

Notice that, out of context, an argument is multivalent also because its form is often misleading and on its own does not prescribe the logic with which it is to be appraised. Consider the following argument whose form is clearly deductive:

(6) John took to drugs because he is poor and poor people take to drugs.

Suppose a teacher wants her students to understand the causes for people taking to drugs. She can put the argument as follows:

(6.1) (It is true that, we know that) poor people take to drugs.
John is poor.
___
(This is why) John has taken to drugs.

In this context and with this function, the argument is intended to be, and is taken as, *proof* or *explanation* of a fact and therefore to be appraised by deductive logic.

Suppose now a social scientist is examining the drug phenomenon in order to understand whether it will involve people like John. She may use the following argument:

(6.2) (It is manifest that) John is poor.
(It has been invariably observed that) poor people take to drugs.
____
(It is reasonable to expect that) John will take to drugs.

In this context the argument can be taken as a *prediction* or *confirmation* of a hypothesis and the logic pertinent to it is inductive logic.

Finally, suppose John is a famous football player and that a politician giving a speech or an interview on television wants to induce the audience to change its mind about our society. He could construct the following argument:

(6.3) You wonder why John has taken to drugs?
Can't you see that John is poor?
In this society, what can the poor do but take to drugs?!

In this case, the argument is an *invitation* to adopt a view or to follow a certain course of action and it cannot be appraised on the basis of its form, only on other substantive views the audience may admit.

As these examples show, the context of an argument provides two kinds of information essential to its classification and appraisal: the *field* of the argument, that is, the concrete, real-life situation in which it is put forward (such as mathematics, physics, politics, etc.) and the *function* of the argument, that is, the specific purpose for which it is advanced in a given field (such as proving a theorem, suggesting a hypothesis, confirming a theory, etc.). Once these elements are provided, arguments can be grouped into such classes as physical explanations, logical proofs, mathematical demonstrations, suggestions of explanatory hypotheses, confirmations of theories, predictions of facts and events, practical advice to follow certain lines of inquiry or courses of action, and so on. Once the classes are identified, arguments can be appraised by a pertinent logic.

Notice that to establish which logic is pertinent to a given argument is not a harmless convention, but a delicate and questionable matter of opportunity. Arguments traditionally ascribed to one logic can be taken away from it and delivered to another. Take proof as an example. According to Aristotle, the proof of an empirical scientific claim is obtained through syllogistic derivation from true premises, so that only deductive logic is pertinent to it. According to modern scientists and methodologists, it is obtained through hypothetico-deductive inference, that is, by using inductive logic. Or take proposals of hypotheses. According to a Baconian, the job is performed by inductive logic; according to a hypothetico-deductivist, such logic does not exist, and what is needed, once the hypothesis has been freely formulated through a psychologically inventive act, is the logic of plausible reasoning or deductive logic alone.

How dialectics taken in both senses actually works in a debate is what now needs to be examined. By way of example, consider two parties: *A,* the proponent, and *B,* the opponent. *A* wants to change *B*'s belief system. Suppose *q* is the belief *A* wants to add to *B*'s system. *A* may advance an argument for *q* that cannot be treated with the tools of formal logic. In this case, *A*'s argument is rhetorically good if it abides by the dialectical rules governing the debate. Alternatively, and more often, *A* may put forward an argument that leads to *q* according to one or other formal logic. In this case, *A*'s job is to find a set of departure-premises *p* accepted by *B* and a bridge-premise *b* also accepted by *B,* linking *p* with *q* according to a logic previously agreed upon. Here *A*'s argument is rhetorically good if these conditions are fulfilled, and *A* is a good arguer if she reasons well and, knowing the subject matter, chooses the right premises.[15]

The set of departure premises are the concessions made by *B* during the debate. But what are bridge-premises? In Aristotle's terminology, they are the *éndoxa,* that is those general premises that are esteemed or considered reputable by competent people. In scientific contexts they are the factors in terms of which scientific debates are conducted and settled. Suppose, for example, *A* puts forward the following deductive argument:

(7.1) The consequences of general relativity are true.
The evolution of the universe is a consequence of general relativity.
———
(It is true that) The universe evolves.

Here the conclusion is drawn from a departure premise ("the evolution of the universe is a consequence of general relativity") together with a bridge-premise ("The consequences of general relativity are true"), referring to a *theory* largely accepted in the field to which the argument belongs (physics). If *B* questions this bridge-premise (for example because, like Bondi, he objects that general relativity has not been tested on a cosmic scale), *A* may resort to a more general one, for example an *assumption.* In this case, *A* may use a deductive argument such as the following:

(7.2) Well-confirmed theories have true consequences.
General relativity is well-confirmed.
———
The consequences of general relativity are true.

Alternatively, *A* may use a different argument and rely on a different bridge-premise, for example a *fact;* in this case the argument may be as follows:

(7.3) If the universe evolves, then there must be traces of a big bang.
Cosmic background radiation has been ascertained.
―――――――――――――――――――――――――――
(It is probable that) The universe evolves.

Of course, most bridge-premises are specific and change from one argumentative field to another (like Aristotle's *proper loci*), while others are more general and typical of most (if not all) scientific contexts (like Aristotle's *common loci*). As there are hosts of them, a detailed list is out of the question. A general framework of the factors involved in scientific debates, however, is needed, and I shall try to provide it.

4.3 The Substantive Basis of Scientific Dialectics

A debate is an exchange of questions and answers between the proponent of an argument and his contenders. The proponent's professed intentions reveal the field and function of the argument in terms of which it can be classified, while the interchange of questions and answers between the proponent and his contenders reveals the factors in terms of which the argument can be appraised.

Reformulating and utilizing to my own ends an expression introduced by Perelman,[16] I shall call these factors the *basis of scientific dialectics* and I shall divide them up into *substantive factors* and *procedural factors*. The former are those substantive notions around which the form of life and culture we call science is organized, and to which one appeals as bridge-premises in a scientific debate; the latter are the rules that govern debates occurring in this form of life. In this section I shall deal with the following substantive factors in scientific dialectics: facts; theories; assumptions; values; commonplaces; presumptions.

One preliminary comment is necessary before proceeding. With regard to these factors, the attitude I adopt is *non*prescriptive. As much as his personal proposals may be appreciated, the philosopher of science is not free to construct systems or models artificially, because he is constrained by the history and practice of science.[17] The substantive factors he has to consider are not his own but those already at work in science. Taken together, they define scientific tradition, that is, a tight web of practices, norms, ways of thought, forms of argumentation, systems of beliefs, and so on, all historically established over a long period of time. For this reason

I shall not be prescriptive, and I shall refer, by way of illustration, to the same sample of scientific rhetoric examined in the previous chapter.

Facts. A typical way of staying on the right side of a scientific debate is to appeal to facts (including those low-level empirical laws that are general factual assertions). Undoubtedly, facts are a vital support, a millstone around the neck of anyone developing arguments that contradict them, but their weight can change according to the various scales on which they are placed.

Consider Galileo. As we have seen in chapter 3.1, when he appeals to facts, at some times he refers to "sensory experience," at others to experience corrected by instruments (telescopic observations), at others to experience resulting from the outcome of experiments, sometimes even to thought experiments. In order for facts to become part of the basis of scientific dialectics and adjudicate the outcome of a debate, criteria of what constitutes a fact must be agreed upon by both sides. Indeed, this is not always enough; two contenders may have the same facts in common but attribute a different weight to them. In this case, the debate will undergo a change and will resort to other substantive factors of dialectics. For example, if a controversy arises over whether or not the moon has seas and mountains, the discussion will soon shift to the reliability of the instrument with which the fact is allegedly ascertained (telescope), and therefore to optical theories.

Theories. As far as theories are concerned, a few distinctions are useful. Since I shall return to the argument in chapter 6, I shall here refer only to *explanatory* theories, i.e., those hypotheses formulated in terms that are not part of what the community considers observational vocabulary but that can be tested empirically. The three theories at the center of the discussion in the previous chapter—the Copernican, natural selection, and big-bang theories—are examples of these.

Like facts, theories too are a heavy millstone in a scientific debate. An argument that shows how a new cognitive claim follows from a well-confirmed and widely accepted explanatory theory is pretty effective. Of course, just how effective it is depends on the kind of link: it is one thing to say a certain thesis "derives from" a theory, another to say it is "compatible with," and still another to say it is "based on" a theory, as in the case of argument (2.1) when Bonnor links the big-bang theory to that of general relativity. Moreover, the argument's efficacy is subordinated to the authority granted to the theory by the audience, as can be seen when Bondi opposes his interlocutors by maintaining that general relativity theory has only been proved on a local and not on a cosmic scale.[18] When doubt is cast on the authority of an invoked theory, the debate again shifts and involves still other factors.

Assumptions. No scientific inquiry is possible without presuppositions. Anyone preparing to delve into natural phenomena must have a preliminary notion of what they are; for example, he must first presuppose that they have an intelligible structure, that this intelligible structure is of a certain type, and so on. Assumptions are precisely these presuppositions. Although they can be considered as past cognitive results taken for granted—in the sense that certain theories (Newton's mechanics, for example), "frozen" into a certain philosophical interpretation (mechanism, for example), can become the presuppositions for further research—assumptions are different from cognitive results. This is evident both from their aim and from their epistemic status.

As regards their aim, assumptions are not intended to provide explanations but rather to interpret the world, that is, to ascribe to it or to specific domains of it a certain structure in terms of a certain fundamental ontology. We could call them *interpretative theories.* As regards their status, interpretative theories, unlike explanatory ones, are not empirically testable assertions but rather metaphysical views that function, to use Kant's terminology, like "maxims of judgment upon which we rely *a priori* in the investigation of nature."[19] The metaphysical core of Lakatos's research programs is a typical interpretative theory in the sense taken here; the same holds for the taxonomical component of Kuhn's paradigms, the ontological element in Laudan's research traditions, and Feyerabend's "natural interpretations."

Interpretative theories can be divided into two classes. *General* substantive assumptions belong to the first class. They are usually considered conditions for the possibility of science as such, and they are highly stable. The most fundamental of them regards the regularity and uniformity of nature, as expressed, for example, by Galileo in the principle *eadem est ratio totius et partium.* One step lower lie assumptions specifying a particular kind of regularity to be expected from nature, such as deterministic causality,[20] or those assumptions expressed by maxims such as *Natura nihil frustra facit,* or *Natura non facit saltus,* or "Nature is simple," etc. The assumptions of the second class could be called substantive *disciplinary* assumptions. They regard a specific portion of the world, usually confined within a single discipline, for example, mechanism in physics or vitalism in biology. The principle of actualism accepted by Darwin is another disciplinary assumption; so is the principle of the circularity of the planets' orbits accepted by Galileo, the perfect cosmological principle considered by Bondi and Gold to be an "assumption that the laws of physics are constant" or as "the only assumption on the basis of which progress is possible without further hypothesis,"[21] and the similar principle of the time-independence of physical laws presented by Lyttleton as a statement of "scientific faith."[22]

Assumptions carry an important weight in a debate. Since they are usually considered conditions for the possibility of science (general assumptions), or of one scientific discipline in particular (disciplinary assumptions), arguments referring to them put forward by the proponent of a thesis can create serious difficulties for the opponent, who may even withdraw from the debate if he realizes he has violated them. Simplicio's strategy when he tries to outplay Salviati by invoking the saying *contra negantes principia non est disputandum*[23] is a good example. Agreement over assumptions must therefore be preliminary; when there is none, it must be obtained by appealing to other substantive factors of scientific dialectics, first and foremost to values. In this sense, the debate between Bondi and his opponents is illuminating: when the latter casts doubts on the perfect cosmological principle, the former resorts to the simplicity of explanations.

Values. Epistemic values can be divided into two classes: empirical science has a *constitutive value,* which is the agreement of cognitive claims with facts, and a set of *regulative values,* such as simplicity, economy, harmony, elegance, falsifiability, high degree of empirical content, fruitfulness, inter-theory consistence, heuristic power, and so on. Unlike regulative values, the constitutive value is sometimes justified by a transcendental argument which claims that without agreement between cognitive claims and facts, empirical science cannot exist. Anyone suspicious of arguments of this kind will anyway admit that the constitutive value is a typical, permanent value of the entire scientific tradition.[24]

The role of values in scientific debate raises three problems. In order for an argument based on values to carry its weight in a debate, it is not enough for the values to be shared; their interpretation must also be agreed upon. Thus, the first problem regards criteria of interpretation.[25] Furthermore, there must be agreement on whether or not, under certain circumstances, such and such a theory or explanation exemplifies such and such a value with such and such an interpretation. Consequently, the second problem regards judgments of exemplification.[26] Not even this is enough, however. Once a given value, criterion, and judgment of exemplification have been accorded, the interlocutors still have to agree on what position that valuc should occupy. Thus, the third problem regards the hierarchy of values.[27] A scientific debate that stretches the limits of values has no effect unless the hierarchies are pinned down in advance. This is precisely the aim of commonplaces of preference.

Commonplaces of preference. According to Aristotle and to the rhetorical tradition, commonplaces are storage tanks or stocks of accepted arguments and opinions at an interlocutor's disposal. Referring to the ones Aristotle called "*loci* of accident," Perelman defines them as "the most general premises, actually more often implied, that play a part in the justification of most of the choices we make."[28] In scientific debates, commonplaces usually function as principles of preference among different values. One of the most common is that of tradition, according to which something that has been repeated often and successfully is preferable to anything new. In our sample of scientific argumentation, we can find examples of this commonplace in Darwin's arguments (6.2) and (6.3) (in favor of the hypothetico-deductive method), in Lyttleton's (6.1) (against the use of the principle of conservation of energy as a definition), and in Bonnor's (6.2) (in favor of the extrapolation on a cosmic scale of general relativity theory). Clearly, though, this commonplace, like any other accepted by a community, works as long as the practice in question is agreed upon and considered authoritative, or until equally significant counter-examples are found to cast doubt on the resemblance between the situations. Commonplaces can take the name of the value to which they accord preference. Thus there may be a commonplace of adequacy ("empirical adequacy is preferable to simple elegance"), of consistency ("internal and external theory consistency is preferable to inconsistency"), of simplicity ("simplicity is preferable to complexity"), etc.[29]

Commonplaces pose some of the same problems as values: first, because they are abstract and general, the problem of interpreting the values to which they refer; second, the problem of making a judgment of exemplification concerning the case for which they have been invoked. A community may agree on a commonplace but disagree over its significance and over the decision regarding whether or not the values it positions hierarchically exemplify the case in question, for example, whether a given theory is really fruitful, etc.

Presumptions. In law, presumptions are suppositions whose effect is to attribute a certain property to an individual because he finds himself in certain circumstances. From a logical standpoint, they have the same form as inferences; for example, "If a child is born of a cohabiting married couple, then the child is the couple's." Their conclusion does not necessarily follow from the premises, but it is nevertheless considered valid until the opposite can be proved; thus the burden of proof rests on the person who casts doubt on it. In science, presumptions can be divided into two types.

The first type are *substantive* presumptions; they refer to facts or explanatory theories and function as guarantees. For example, "If R is a serious researcher, then the experiments and observations made by R are reliable." A presumption of this kind was often invoked in favor of Tycho Brahe's data which Kepler and even Galileo generally considered correct. Or again, "If a law L or a theory T is well confirmed, then it is firmly grounded (and it cannot be violated)". As we have seen, a similar presumption regards the law of energy conservation and is invoked against the steady-state theory.

Presumptions of the second type are *regulative;* they refer to assumptions and work (like commonplaces) as principles of preference. For example, "If nature has a mathematical structure, then mathematized theories are to be preferred to others," a presumption Galileo relies on; or again, "If the causes at work in the world are the same as the ones we see at present, then actualistic theories are to be preferred to transcendental ones," a presumption to which Darwin refers.[30]

Presumptions are not absolute truths; they are true until they are proved to the contrary. In this respect they are like facts and theories, which are also subject to revision and are valid only until someone comes along and denies or rejects them. There is, however, one basic difference. In a debate, he who upholds a claim by grounding it on a fact or a theory must bear the burden of proof; on the other hand, he who covers a claim with a presumption shifts the burden to the other side. Thus presumptions carry significant weight in a final point-count or in adjudicating victory in a debate.

Having exhausted the list of substantive factors of scientific dialectics, I have to say something about their order. The one I have given here does not presuppose any hierarchy of levels. In other words, my intention is not to claim that a scientific debate starts off with facts and proceeds, level by level, to presumptions. Nor do I maintain that divergence at one level can be reconciled at a higher level. Indeed, a scientific debate can involve many factors simultaneously and divergence on one factor can be reconciled by appealing to any other. In general, there is no such thing as a privileged departure point and a guaranteed point of arrival (even the most consolidated presumptions can be doubted). The departure point is established, case by case, by the participants in the debate, while the arrival point is simply where, case by case, the debate comes to a halt. Nor is there a predetermined path between these two points. Once the participants have agreed on the former, how to get to the latter also depends on the circumstances of the debate. From this point of view, the position maintained here resembles the one Laudan calls a "reticulated model of justification," in which "axiology, methodology, and factual claims are inevitably intertwined in relations of mutual dependency."[31] What is missing here, obviously, is methodology. Considered in the sense of arbiter of the scientific game, methodology is substituted by substantive and procedural factors. Considered, rather, in its specific sense of a set of rules of decisions (the standard rules of acceptance, rejection, and preference), methodology is absorbed by values. A rule of decision (for example, the rule "Prefer the theory that does not make use of ad hoc hypotheses") is an imperative depending on an epistemic value (for example, the value of falsifiability).[32]

4.4. The Appraisal of Scientific Arguments

On the basis of substantive factors, the main valuative notions of scientific argumentation can now be introduced. I shall try to offer explications which are able to satisfy the two standard requirements of adequacy and precision.

The first notion is *pertinence*. As a matter of fact, pertinence is an ambiguous concept, both descriptive and valuative. Saying an argument is pertinent is a bit like saying a person is intelligent: this epithet can be as much a description of fact (for example, that person's reaction to an IQ test), as a judgment laden with moral and social consequences. Only use can dissolve the ambiguity. I thus propose the following explication:

Expl.1

A scientific argument in a given field and for a given function is *pertinent* if the reasons supporting its conclusion belong to the substantive factors of scientific dialectics admitted in that field and for that function.

This explication is no doubt adequate, for pertinence is relative to a field. For example, an argument trying to conclude that a certain astronomical hypothesis is true or probable because it is based on Holy Scripture would not be considered pertinent because it is not based on admitted substantive factors, first and foremost, empirical testability. Pertinence is also relative to function. For example, arguments of the possible (such as Darwin's) would not be considered pertinent if they were implemented with a more ambitious aim than merely suggesting a hypothesis be taken into consideration. It is evident, though—and this is true for all the other explications—that Explication 1 can be considered precise only if field and function of the argument are clearly specified. When a field is redefined and its limits are stretched or shrunk, Explication 1 can become vague and the question of whether or not an argument is pertinent can give rise to controversy. Typical examples are arguments taken from religion in the fields of astronomy and cosmology.

After pertinence we have *validity*, for which I propose the following explication:

Expl.2

> A scientific argument in a given field and for a given function is *valid* if in favor of its conclusion a winning dialectic strategy exists on the basis of the substantive factors of scientific dialectics.[33]

There is little to be said for now about this explication. Intuitively, it appears to be both adequate and operative; but a pondered opinion cannot be expressed until we know more about the winning dialectical strategy.[34]

After validity comes *strength*. Perelman says quite rightly that strength is connected to both validity and efficiency[35] and "depends considerably on a traditional context."[36] We can try to solve the difficulties that arise from this connection if we consider that according to Expl.2 validity is an absolute notion.[37] If an argument is valid (in the sense that there *is* a winning dialectical strategy for its conclusion), then it is valid independently of the circumstances, or it is valid in relation to the universal audience, which amounts to the same thing.[38] Strength is a relative notion. If an argument is strong, it is strong not because it "must convince every reasonable mind,"[39] but because there is a winning dialectical strategy for its conclusion *within* the situation in which it is advanced, and because it must convince every reasonable mind that admits the specific factors relevant to that situation.

Let us call a *dialectical situation* (s) the state of a scientific debate at a given moment, and the *configuration of the substantive basis of scientific dialectics* (c) the arrangement of substantive factors in force holding in a given situation. I can then propose the following explication:

Expl.3

> A scientific argument in a given field for a given function is *strong* if in favor of its conclusion a winning dialectical strategy exists on the basis of both the premises conceded in the dialectical situation and the configuration of the substantive factors of scientific dialectics in force in that situation.

Unlike validity, strength admits of degrees. This too can be explicated.

Suppose a situation s_1 and a configuration c_1 first. Given two arguments A and B, if there is a winning dialectical strategy for A on the basis of c_1, then A is strong. But such a strategy may not be found; there may be two dialectical strategies, one for A and another for B, such that neither is able to subdue the other. Argument A may be stronger than B if the strategy for the conclusion of A appeals to more factors or if the factors on which it rests are considered by the community to be more important. This was the case in the debate between the big-bang and steady-state theories in the fifties. Arguments supporting the former happened to be stronger than those favoring the latter, mainly because of the weight attributed to the law of the conservation of energy.

Suppose now a change from s_1 to s_2 takes place, accompanied by a change from c_1 to c_2. In this case an argument *A* may be stronger or weaker in s_2 because c_2 alters the previous configuration. An example is given by some of Darwin's arguments. His resistance to the idea of publishing the *Essay* containing the theory of natural selection in 1844 was overcome in 1856 when the scientific community changed its mind about one of the basic assumptions of the theory, that is, transformism. The same arguments that could once have been considered weak were thus strengthened.

The following explications can therefore be proposed:

Expl.4

A scientific argument *A* in a given field and for a given function is ***stronger*** than an argument *B* in the same dialectical situation if the dialectical strategy in favor of the conclusion of *A* is based on more substantive factors or on more important substantive factors than those contemplated by the dialectical strategy favoring the conclusion of *B*.

Expl.5

A scientific argument *A* is ***stronger*** in a dialectical situation s_2 than in another s_1 if, during the change from s_1 to s_2, the dialectical strategy in favor of *A* becomes winning or if it relies on more, or more important, factors.

Finally, comes ***efficiency***. One can say an argument is efficient if it manages to convince, but an argument that convinces may convince some and not others, or some more than others. Here, the notions of situation and configuration come to our aid. I suggest the following explication:

Expl.6

An argument in a certain situation is ***efficient*** for an interlocutor (or an audience) *I* if the reasons adduced in support of its conclusion belong to the configuration of substantive factors of scientific dialectics that *I* considers optimal in that situation.

Intuitively, this explication is adequate. Suppose an individual or a group of individuals maintain that a certain scientific explanation, prediction, experiment, technique, etc. must fulfil precise requirements before it is acceptable; it is natural that they will find an argument that appeals to these requirements to be most convincing. The explication is also operative because knowing what configuration is accepted by *I* allows one to pin down precisely whether or not a certain argument will be efficient for *I*. Explication 6 also seems to explain why an efficient argument is not necessarily strong. An argument is considered efficient on the grounds of the configuration of factors accepted by *I*, while it is considered strong on the grounds of the configuration in force in the situation in which it is introduced. If the two configurations overlap, an efficient argument is also strong. However, they can also not overlap.

We must now clear up the essential point we left aside earlier. In order to be operative, most of our explications need the notion of winning dialectic strategy to be defined. I propose the following explication:

Expl.7

> A dialectical strategy in favor of a scientific thesis *T* is winning for one side *P* against the other *Q* if, on the basis of the rules that govern scientific debates, *P*, starting with the premises granted by *Q* and with the substantive factors of scientific dialectics, forces *Q* either to assent to *T*, or to stay silent, or to withdraw from the debate.[40]

All we need do now is to specify the rules that govern scientific debates, that is, what I have called the procedural factors of scientific dialectics.

4.5 The Procedural Basis of Scientific Dialectics

As far as these factors are concerned, again, my aim is not prescriptive. I do not intend to construct a system of formal dialectics or a dialectical game, but to make explicit those rules that are provided by scientific practice found in the history of science and accredited by scientific tradition.[41] There are two kinds of rules that apply to the procedural factors of scientific dialectics: rules for conducting a debate and rules for adjudicating a debate.

Rules for conducting a debate. These rules discipline the type of exchange allowed between interlocutors, that is, they establish admitted moves and counter-moves. In many respects, a scientific debate is no different from a normal debate. However, not all debates are equal. For example, while some would consent to an interlocutor withdrawing a thesis he himself has previously admitted, others would not. A scientific debate cannot be regulated by rigid rules of this kind. It is current practice for a scientist supporting a certain thesis (e.g., a theory) on the grounds of certain reasons (e.g., a set of empirical evidence) to withdraw at least one of these reasons without feeling automatically obliged to withdraw the theory he is supporting as well.

A few typical moves and counter-moves are listed below; the list makes no claim to completeness.[42]

Moves	Counter-Moves
(1) S	(1′) a. S
	b. $\neg S$
	c. S?
(2) S / r_1	(2′) a. $\neg S / r_1$
	b. $\neg S / \neg r_1$
	c. $\neg S / r_1 + r_2$
	d. $S' / r_1 + r_2$
(3) $- S$	

With move (1) of this list, the proponent launches the debate by saying "I state *S*." The answer (1′) can be of three types: (1′a) "I admit *S*," (1′b) "I deny *S*," (1′c) "I require reasons for *S*." If the proponent provides the reasons requested (move 2), the interlocutor has four counter-moves open to him. Firstly, he can reply (2′a): "I deny *S* because r_1 is a reason for not-*S*"; in this case he is casting doubt on the inferential link between the proponent's thesis and the reasons adduced by him. Alternatively, he could use (2′b) "I deny *S* because I refute r_1"; in this case he is rejecting the proponent's thesis by denying the reasons adduced. Furthermore, he could resort to (2′c) "I deny *S* on the grounds of another reason that can be added to the reasons you adduce"; in this way he makes the proponent consider other reasons and, if these reasons have already been admitted by the proponent in the debate, he challenges him to prove that his main thesis is compatible with them. Finally, the interlocutor can object (2′d) "I state *S*′ on the grounds of another reason that can be added to the one you adduce"; this is a typical counterattack. Move (3) is equivalent to "I withdraw *S*" and is always conceded. If the proponent withdraws the thesis that is the object of the argument, he obviously has to retire from the debate altogether; if he modifies it, then the debate shifts to another thesis; if he withdraws another thesis previously admitted (for example, a reason adduced), then he must prove that the residual reasons are enough to defend the main thesis.

For how long can a debate governed by these rules go on? Until one of the two interlocutors is overcome by his adversary's arguments. Precisely when is determined by the rules for adjudicating a debate.

Rules for adjudicating a debate. These rules determine the points bestowed on each side and award the final victory. From a *logical* point of view, a scientific debate between two sides *A* and *B* is adjudicated in favor of *A* when *A* confutes *B*. Since, as Aristotle wrote, "a refutation is a deduction to the contradictory of the given conclusion,"[43] if B affirms a thesis T and then, in the course of the debate, is obliged to accept $\neg T$ (because, for example, he accepts O and $O \rightarrow \neg T$), then B is beaten and A scores a victory.

Substantive factors of scientific dialectics play an essential role here. The logical *strategy* for confuting a thesis consists in finding one or more concessions made by the interlocutor which, united with a shared substantive factor that acts as a bridge-premise, leads to the negation of that thesis. Let us imagine a debate between a proponent maintaining a thesis K and an opponent who negates it. C is a premise belonging to the subset of premises granted by the opponent (concessions) which is included in the set of accepted premises P, while F is one of the substantive factors of scientific dialectics. Thus the proponent's strategy for confuting his opponent consists in deriving K from C and F, as shown in figure 4.1.[44]

As one can see, the technical role of substantive factors here is the same as that in Aristotle's *Topics*. Their function is to provide a stock of premises, starting points, or support bases for disputants in a debate.[45] In this sense, they resemble Toulmin's "warrants of argumentation," or—to borrow a terminology used for different aims—"covering laws" or "license tickets."

Note that while the strategy is always the same, *tactics* can vary. Instead of the sequence *CFK* ("*C, F;* therefore *K*"), the proponent may choose the sequence *CKF* ("*C;* therefore *K,* because *F*") or the sequence *KFC* ("*K;* because *F,* given *C*") or the sequence *FCK* ("*F;* therefore *K,* given *C*"). It is up to the proponent's perspicacity and cunning, his knowledge of the subject matter, and the commitments of the opponents, to find the sequence that is most suitable. At times he may find it more convincing to state the bridge-premise explicitly at the beginning of the debate; at others he may find it more satisfactory to start from the concessions, present his conclusion, and finally introduce the bridge-premise; at still other times, he may find it easier to anticipate the conclusion and insert the bridge-premise when an appropriate concession is made. More often, he can decide to leave the bridge-premise implicit, either because he considers it

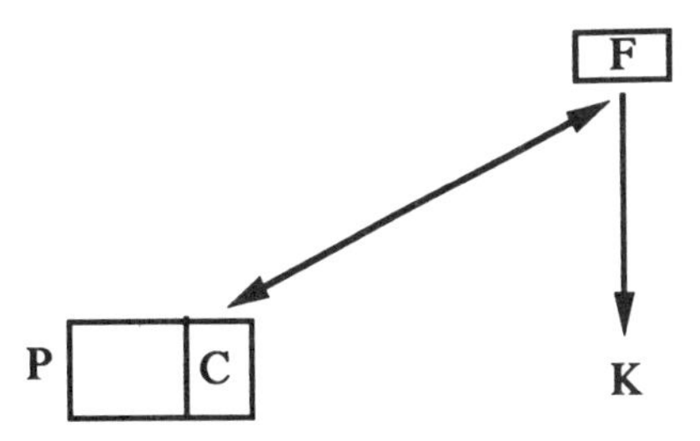

Fig. 4.1.
A strategy for confuting an opponent.

obvious or because he does not want to extend the debate to other related points; in this case the argument takes on an elliptical form.[46]

Although a confutation strategy aims at a definitive (logical) knockout, in practice things are different. A proponent rarely maintains only one thesis; he is more likely to uphold a set of theses, thus leaving himself ample room for maneuver if he should fall victim to confutation. Alternatively, an opponent may not make all his commitments explicit, or he may object that his view has been misunderstood. Alongside the logical confutation rule, then, we must consider other ways of adjudicating a debate. From a *pragmatic* point of view, scientific practice shows a debate between *A* and *B* is adjudicated in *A*'s favor under one of the following circumstances:

(C1) *B* does not offer reasons in support of his thesis belonging to the admitted substantive basis.

(C2) *B*, who has the burden of proof, shifts it to *A*.

(C3) *B* does not answer the problems he himself recognizes as relevant during the debate.

(C4) *B* contradicts a thesis previously admitted, presupposed, or derived from one or another of his concessions and cannot settle the contradiction.

(C5) *B* denies one or another of the substantive factors in the shared configuration of the basis of scientific dialectics.

(C6) *B* denies a presumption he himself accepts.

(C7) *B* is led to affirm a thesis contrary to an accepted presumption.

(C8) *A* proves his own thesis starting with one of *B*'s concessions.

A few examples will be useful to illustrate these rules.

(C1) In Galileo's *Dialogue,* this circumstance can be found in the part of the debate regarding the perfection of the world that Simplicio derives from the perfection of the number three, to which Salviati objects that "it would have been better for him to leave these subtleties to the rhetoricians, and to prove his point by rigorous demonstrations such as are suitable to make in the demonstrative sciences."[47] Here Simplicio is accused of bringing arguments that are not relevant to mathematics.[48]

(C2) The burden of proof rests on the person who at the beginning, or in the course of the debate,[49] advances a cognitive claim or opposes an accepted claim. An example of the shifting of the burden of proof is provided by *ad ignorantiam* arguments such as "If you do not prove the truth (falsehood) of *p,* then *p* is false (true)." Bondi's argument in favor of the perfect cosmological principle, which states that if this principle is not admissible then cosmology is "a very much more difficult subject than I would like to tackle"[50] or "is no longer a science,"[51] does not violate this rule because its intention is not to conclude that such a principle is true.[52]

An example of the shifting of the burden of proof can be found in Simplicio's resorting to *ipse dixit,* to which Salviati retorts that "Aristotle acquired his great authority only because of the strength of his proofs and the profundity of his arguments."[53] Or again, in Simplicio's explanation of downward movement on earth — "the cause of this effect is well known; everybody is well aware that it is gravity" — to which Salviati replies: "You are wrong, Simplicio; what you ought to say is that everyone knows that it is called 'gravity.'"[54]

(C3) The problems referred to here can be either internal, that is, concerning the formal structure of a theory and its relation to the observational domain, or external, that is, concerning the relations between that theory and other theories or certain assumptions.[55] Many examples can be adduced. One regarding internal problems can be found in the controversy over Galvani's theory of "animal electricity" and Volta's theory of "contact electricity." In 1794, Volta challenged Galvani to produce signs of electricity without using metal arcs between the nerve and the muscle of a frog. Galvani and his followers accepted the challenge and succeed in obtaining these signs. Having considered the problem crucial, hinging the destiny of his own theory of metallic contact electricity on a negative outcome, Volta was forced to withdraw his theory. The palm of victory, at least temporarily, changed hands.

(C4) This situation is illustrated in the controversy between Galileo and Father Scheiner concerning sun spots that I shall be examining in the next chapter. Note that (C4) does not oblige *B* to withdraw from the debate immediately. He can also retrace his steps, change or reinterpret his thesis, as long as these moves do not signify withdrawing the main thesis. However, (C4) does oblige *B* to settle the contradiction. The less successful he is in doing so, the weaker his thesis becomes. This is frequently the case in Darwin. For example, as we can recall from his argument (9.3), when dealing with one objection to his theory Darwin maintains that there must be a cause for each and every slight individual difference, to the extent that "if the unknown causes were to act persistently, it is almost certain that all the individuals of the species would be similarly modified." This concession weakens Darwin's position because it makes the natural selection of variations dependent on specific, not blind, forces.

(C5) Whoever is unfortunate enough to find himself in this situation is usually knocked out, but he can always get up and resume the fight because the configuration of factors of scientific dialectics is not established once and for all; it can itself become the object of discussion. Thus, in the cosmological debate, a typical move for critics of the steady-state theory is to object that this theory contradicts certain admitted facts, laws, and values — in particular the principle of the conservation of energy — and then to claim that their own configuration of these factors is universally accepted. An equally typical counter-move for upholders of the same theory is to retort that the contradiction is not fatal because still another configuration that saves their own theory is possible. Clearly, though, if a configuration c_1 that favors theory T_1 is generally accepted, while a different configuration c_2 that favors T_2 is not supported by good reasons at least partially independent of T_2, then upholders of T_2 find themselves in serious difficulty.

(C6) and (C7) These situations are extremely difficult because presumptions are among the most decisive factors in the configuration of the substantive basis of dialectics. An interlocutor led to recognize that his own theory violates a presumption he himself has admitted while his adversary's theory is compatible with it, will almost certainly lose the debate. Galileo makes use of this sort of confutation several times, as when he places the presumption of simplicity in Simplicio's mouth and then proceeds to highlight the "monstrous chimera"[56] of Ptolemy's theory against the harmonic nature of Copernicus' theory. The technique is again to turn an enemy's own weapons against him.

(C8) This is perhaps the most desperate situation, since whoever finds himself in it is forced either to withdraw or modify his thesis. A few critics of Darwin adopt this technique. Jenkin, for example, stresses that precisely the Darwinian mechanism of selection of variations together with blended inheritance of characters does not lead to the conclusion of gradual evolution but to that of the creation "from time to time" of the species.

The reconstruction of scientific dialectics at this point may be considered complete; it could be objected, however, that it is not precise enough to award certain, definitive victories. The answer is both yes and no.

Yes. Like Freud's analysis, a scientific controversy is in principle *interminable.* Even when one side is in serious difficulty, it can find a way of getting out of the corner and making a counterattack, thus turning the whole situation around in its favor. There is a technical reason for this which has already been mentioned. Suppose that at a certain point of a debate in the column of a proponent A we find the sequence:

T_1

. . .

. . .

$\neg T_1$

In this case, A would affirm two *contradictory* theses and would be confuted. This situation is fairly rare, however. What usually takes place in practice is that a column looks like this:

$$T_1, T_2, T_3$$
$$\ldots$$
$$\ldots$$
$$\neg T_4, \neg T_5, \neg T_6$$

Suppose now the following implication holds among these theses:

$$\{T_1, T_2, T_3\} \dashv \{T_4, T_5, T_6\}.$$

Call this implication a *dialectical implication*, which means that at least one thesis in the first set is false and at least one thesis in the second set is true.[57] Thus A affirms two sets of theses that are *incompatible*. But this does not mean that B has won the debate. Incompatibility is not like contradiction; it can be solved. For example, A can abandon one of its theses and withdraw, say, to the position $\{T_1, T_2 \neg T_3\} \wedge \{\neg T_4, \neg T_5, \neg T_6\}$. If the abandoned thesis is not the one to which A was explicitly committed as his main thesis, then the debate can go on.

No. Like Freudian analysis, a scientific debate is in practice *terminable*. After Harvey, it would be rather difficult to go back and reopen the controversy over the circulation of blood, just as after Galileo it would be hard to rekindle a debate on the physics of natural places. Feyerabend has castigated methodology with the argument that if methodological criteria do not go hand in hand with temporal limits they are mere "verbal ornaments."[58] However, this argument is not valid against dialectics. One could say that the point at which a debate is settled, and beyond which there is no point in venturing, is reached when an interlocutor in difficulties on the basis of the substantive and procedural factors of scientific dialectics is at a loss to find adequate counterattacks and simply repeats his own arguments or ignores his adversary's arguments.

It could also be objected that scientific dialectics, inasmuch as it is logic, is a canon of appraisal exactly as methodology intended to be and consequently that, at the most, it either widens or relaxes methodology. Here too the answer is yes and no.

Yes. Scientific dialectics does in fact widen and relax methodology. It widens it because it takes into consideration many more factors than those contemplated by the typical set of methodological standards. It relaxes methodology, at least in the sense suggested by Hempel,[59] because, while taking research away from the Cartesian "certain and simple rules," dialectics does not make methodology dependent on "purely idiosyncratic individual factors."[60]

No. Scientific dialectics overcomes methodology and it is not a "supplement" of (deductive or inductive) logic.[61] Rather, it is the logic of scientific discourse, the logic of belief change in scientific debate.[62]

Five

The Dialectical Model of Science

On every issue there are two arguments opposed to each other.

Protagoras, in Diogenes Laertius, *Lives of Eminent Philosophers*

From its *seeming* to me—or to everyone—to do so, it doesn't follow that it *is* so. What we can ask is whether it can make sense to doubt it.

L. Wittgenstein, *On Certainty*

5.1 Science: A Game with Three Players

From a historical point of view, what we today call methodology was first developed between the sixteenth and seventeenth centuries thanks to the introduction of a very simple idea, one of those simple ideas that then go on to alter radically the face of the earth. The idea amounts to this: if you want to know more about nature, then look at it and stop quarreling with people who talk about nature without observing it, or who take no more than a brief glance at it.

The background to this revolutionary discovery is very interesting because of its similarities with the contemporary context in the philosophy of science.

The founding fathers of modern science (the Fathers, for short), tried to find, as Bacon put it, a passage between "the presumption of pronouncing on everything and the despair of comprehending anything,"[1] or as Gassendi repeated, "a way between skeptics and dogmatics."[2] The solution they hit on is well known: against the dogmatism of scholastics they claimed that natural science does not progress through deduction from evident principles but through induction from observations and experiments; and against the relativism of skeptics they maintained that scientific knowledge, although not certain, is well founded and reliable within its own limits. As Gassendi wrote, "If it is permitted to know many things, they are never such that they can be discovered following the rules of that famous Aristotelian science; rather the only way is through experience or according to appearances."[3] Put schematically, the passage that was found was a new "discourse on method."

Dialectics was the first to be abandoned. By way of example, let us see a few typical reactions, mingling antischolastic sentiment and direct criticism.

Bacon claimed that the aim of science is "to overcome, not an adversary in argument, but nature in action."[4] Only sciences based on opinion and dogmas can make use of dialectics, "for in them the object is to command assent to the proposition, not to master the thing."[5] Thus Bacon criticized Aristotle, ascribing faults to him that were not his, because "with the dialectical arguments attributable to him (about which he boldly boasts) he has corrupted natural philosophy."[6] Galileo's attitude to Aristotle was more balanced,[7] but he was no less polemical towards scholastics and dialecticians. He accused them of futile "altercations,"[8] objecting that "this kind of person thinks that nature is a book like the *Aeneid* or the *Odyssey,* and that truth should not be sought in nature or in the world but (in their very own words) in a comparison with the texts."[9] Descartes wrote that "ordinary dialectic is of no use whatever to those who wish to investigate the truth of things. Its sole advantage is that it sometimes enables us to explain to others arguments which are already known. It should therefore be transferred from philosophy to rhetoric."[10] Robert Hooke wrote that "the Science of Nature has already been too long made only of the *Brain* and the *Fancy:* it is now high time that it should return to the plainness and soundness of *observations* on *material* and *obvious* things."[11] Similarly, Thomas Sprat claimed that the members of the Royal Society "have indeavor'd to separate the knowledge of *Nature* from the colors of *Rhetorick,* the devices of *Fancy,* or the delightful deceit of *Fables.*"[12] Towards the middle of the seventeenth century, Hobbes countered logic with rhetoric, maintaining that "the end of that is truth, of this victory,"[13] and, at its close, Locke took up the same theme, writing that these generic maxims that are resorted to in dialectics are "of great use to stop wranglers in disputes, but of little use to the discovery of truth."[14]

The value of the antischolastic controversy that gave rise to modern science cannot be underestimated; and it would be futile to applaud it once again were it not to remind those who seem to have forgotten that at the time the Fathers made a useful *choice,* took a *decision* that, as far as the cognitive and technical results were concerned, proved to be fruitful.[15] However, we must recognize that, as regards the philosophical image of science, the decision taken had negative consequences. There is no doubt that there were valid reasons for opposing the scholastics, reduced to "creating a world out of categories," as Bacon wrote,[16] but associating antischolasticism with antidialectics was a big philosophical mistake: it must be admitted that on this point the son of the physician at Macedonia's court had greater wisdom than Queen Elizabeth's lord chancellor.

Bringing dialectics into science is not just a matter of making small adjustments here and there; the Fathers' very image of science is irrevocably altered.

Consider this view. What the Fathers had in mind can be represented as a game with two players. In this game, one player (the inquiring mind, *I*) asks the other player (nature, *N*) questions through observations and experiments; the latter provides answers through data and results. The match is over when *I* has forced *N* to reveal his secrets. To know when that has been achieved, however, certain rules are needed. This is the task of method, *M,* which dictates rules for each step required by the procedure that links *I* and *N*. For example, if the procedure is hypothetico-deductive, there must be rules regarding what counts as a good observation, a legitimate hypothesis, a severe test, a genuine confirmation or falsification, etc.; or else—as is the case with current, sophisticated methodologies—there must be rules regarding whether one theory is better than another, whether a research program is progressive or regressive, whether one tradition is preferable to another, etc. The layout of the whole game with two players can be presented as in figure 5.1.

In this figure, method is depicted as an eye, because it is precisely like God's eye; from its privileged vantage-point, truth and error are infallibly revealed. Descartes made this point clear; his Rule IV claims that "we need a method if we are to investigate the truth of things,"[17] such that if one painstakingly follows the rules of this method, "one will never take what is false to be true." Galileo was of the same opinion. He argued that if one

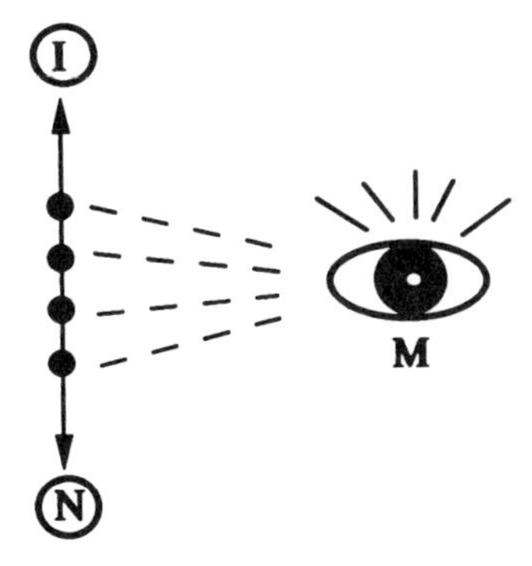

Fig. 5.1.
The methodological model of science.

scrupulously upholds the method of "sensory experiences and necessary demonstrations," one will acquire knowledge equal to God's, if not *extensively,* at least *intensively.* Of course, dialectics and rhetoric are strictly forbidden to partake of this method. Descartes was explicit on this point, too. He wrote that method is complete if it explains the correct use of the only two operations required of knowledge, that is, intuition and deduction: "as for other mental operations which dialectic claims to direct with the help of those already mentioned, they are of no use, or rather should be reckoned a positive hindrance."[18]

Due to the role of arbiter and judge played by method, I shall call this image of science the *methodological model.* One of Goethe's aphorisms, paraphrasing the Gospel,[19] sums it up well: in this model "nature, under torture, is silenced; her honest answer to an honest question is: Yes, yes; no, no! Anything else comes from the Evil One."[20]

Obviously, different presuppositions underlie the methodological model. In particular: that nature provides data (or clear and distinct ideas) through observations (or intuition) and experiments; that data relevant to the confirmation or falsification of hypotheses or theories are independent of them; that data constitute the only significant test of our cognitive claims; and finally that both the procedure linking data to cognitive claims and the methodological rules underlying the procedure are universal, i.e., do not change when theories change.

It is well known, as we recalled in the Introduction with a philosophical fable, that in contemporary philosophy of science, all these presuppositions have one by one fallen by the wayside due to a number of results obtained from sources ranging from epistemology and philosophy of language to logic and the history of science. Nowadays it is platitudinous to maintain that data are theory-laden, that there is no logic of discovery leading from data to cognitive claims, that there is no clear distinction between observational and theoretical concepts, that theories cannot be reduced to their empirical basis, that they are underdetermined by it, and finally that there is no universal method. The outcome of these results, combined with the tendency to extremism proper to all revolutions (including those in philosophy) and, above all, with what I have called the "Cartesian syndrome," is that the methodological model has been overturned into what may be called the *counter-methodological model.*

In this model the arbiter has been eliminated with the result that the correctness of the game—that is, the rationality of science—has vanished too. Nonetheless, there are still only two players, the same ones as before: nature on one side and the inquiring mind of the scientist on the other. The only difference is that nature's voice (that is, the empirical basis) is softer, while, to go back to Goethe's metaphor, the Evil One (for example,

Feyerabend's "external factors," or Lakatos's "mob psychology") makes such a din that the inquiring mind is obliged to shout louder, until nature's voice is overwhelmed and reduced to meaning no more than the private interests, tastes, or idiosyncracies of those who inquire into it.

Note the historical analogy. The shift from the methodological to the counter-methodological model in a significant portion of today's philosophy of science is similar to the one from dogmatism to skepticism in the sixteenth and the seventeenth centuries. The positivistic dogmatism prevalent in the first half of the twentieth century, according to which science is knowledge with solid foundations, stands to the scholastic dogmatism of the past, which stated that science is knowledge based on primary principles, just as recent relativism, according to which science is no more or less rational than any other intellectual venture, stands to the skepticism of the past, which decreed the impossibility of science. Modern positivists have replaced the scholastic *ipse dixit* with expressions such as *natura dicit, experimenta probant,* while today's relativists substitute Francisco Sanches's *nihil scitur* for Feyerabend's "anything goes."

If we now consider dialectics rather than method as the logic of science, the whole image changes because of the essential, constitutive role played by interlocutors. Due to this role, science becomes a game with *three* players: an inquiring mind, or, more realistically, a group of the community C_1, nature N, and another group of the community C_2. In this game, C_1 opens the match, advancing a question, a problem, a hypothesis h_1 and supporting it with observations or experimental outcomes O, N provides data e, C_2 by advancing h_2 discusses both h_1 and e, and a debate D takes place between C_1 and C_2 in terms of the factors F of scientific dialectics. The match is over when C_1 and C_2 have come to an agreement about which solution is acceptable. Schematically, the situation is as in figure 5.2.

I shall call this the *dialectical model* because of the role played in it by the debate among interlocutors, and therefore by the dialectical techniques of confutation and persuasion. What is immediately clear is that in this game there is one more player, the community, but one less protagonist, method. The community cannot be considered an arbiter or supervisor, for it is a player. The lack of an arbiter, however, does not bring about the same disastrous effects as in the counter-methodological model because the debate between C_1 and C_2 regarding the solution about N is regulated and constrained by D, the factors of scientific dialectics. To return once again to Goethe's metaphor, in the game with two players an arbiter is needed. When, due to the Evil One's bedlam, the arbiter disappears, the game degenerates and one player overwhelms the other. In the game with three players, the arbiter also vanishes, but so does the Evil One.

Let us take this point further. In the methodological model, nature,

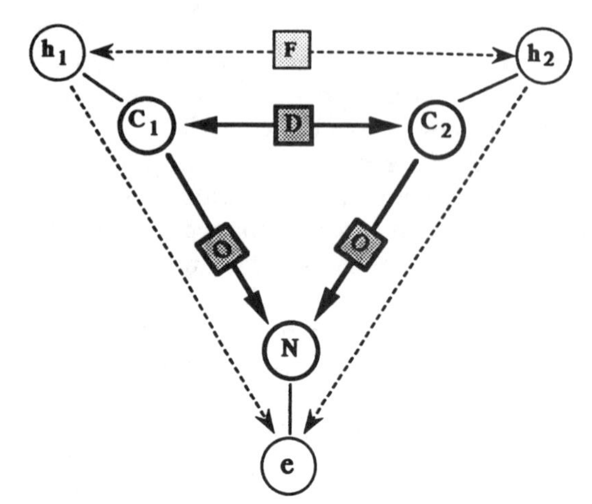

Fig. 5.2.
The dialectical model of science.

forced by the rules of method, speaks clearly and the inquiring minds *record* its real voice. Hence Descartes says that if one follows the rules, "one will never take what is false to be true." In the counter-methodological model, nature utters sounds and the private interests or idiosyncracies of its investigators *construct* a possible meaning from the sounds. Hence science becomes mere "social convention." In the dialectical model, nature reacts and scientists *agree upon* its correct answer through a debate based on the factors of scientific dialectics. Agreeing upon a correct answer means neither passively listening (or reading), nor fabricating under the pressure of private or social interests: it means, rather, finding that view (cognitive claim) that best holds out against criticism. Nor does it mean agreeing upon a mere conversational or hermeneutic matter, because the agreement has no value if it does not respect the constraint imposed on the debate by empirical evidence and facts which are among the factors of scientific dialectics. The Fathers were right when they criticized the scholastics for not heeding this constraint, but they were wrong in visiting the sins of the children on their father. Aristotle had always pointed out that without experience there is no natural science.[21] He criticized those who had forgotten this golden principle,[22] and noted that confuting rival theses (through dialectics) with "the ability to puzzle on both sides of a subject" and using persuasive means (through rhetoric) "on opposite sides of a question" are not the same as a mere verbal discussion of opinions at hand since this involves a critical analysis of *all* the theses and objections *proper* to the subject itself.[23]

Considered in this light, the dialectical model offers a promising solution to the contrast between internal and external factors, a dichotomy Kuhn has called "the greatest challenge now faced by the profession."[24]

The factors of scientific dialectics, indeed, are neither internal nor external: they are simply *proper* to science. Alternatively, they are both internal and external, because on the one hand they belong to science and nothing else, and on the other they can undergo transformations along with the transformations of culture.

Changing thus our image of science, the dialectical model also ascribes new duties to the philosopher. In the game with two players, science is solidly *grounded* knowledge and the philosophy of science has the task of establishing the logical links that bind a cognitive claim *h,* advanced by *I,* to empirical evidence *e* provided by *N,* for example, by calculating the degree of confirmation *c* (*h, e*), as well as of establishing the rules that govern decisions concerning *h* and *e,* for example, by adding a "methodological supplement" to deductive logic[25] or a "methodology of inductive logic."[26] (When the game with two players degenerates, the *h–e* link becomes irrelevant, one obtains "science without experience,"[27] and the philosopher becomes at the most an anthropologist.) In the game with three players, science is always *transformation* of previous knowledge, and the philosophy of science has the task of analyzing the transformation's dynamics. Here the final *h* is neither derived from nor superimposed on *e;* it is the result of a selective competition with other hypotheses advanced by the community or present in the community's accepted knowledge.

We seem to be in the fortunate position of being able to take advantage of extreme views without suffering too many inconveniences. On the one hand, the dialectical model does not rid science of the traditional empirical factors of appraisal, nor does it neglect the so-called pragmatic factors; it simply brings them *within* the framework of concrete scientific discussions. Thus the weight of facts, stressed by Popper and Lakatos, does not vanish but is added to that of other elements;[28] while psychological and sociological factors, considered vital by Feyerabend and sociologists of science, operate through the dialectical *filter* that regulates scientific debate.[29] On the other hand, the dialectical model assumes that the formation of scientific consensus is conversational but discloses the constraints of these discussions. The basis of scientific dialectics makes Rorty's ideas of "routine conversation" and "good epistemic manners" more specific and better defined.

Judged from the viewpoint of the dialectical model, both sociology and hermeneutics of science reveal similar defects. The sociology of science bypasses the factors of scientific dialectics altogether and links social and cultural conditions directly to scientific products.[30] But this link is premature. To cite two examples, Victorian culture in England could have influenced the genesis of the theory of natural selection; likewise, with the intellectual atmosphere under the Weimar republic and the birth of quantum theory. But such influences are not direct; they are mediated by scientific dialectics. Take these factors away and the sociology of science soon becomes a master key for any lock. The hermeneutics of science also bypasses the factors of scientific dialectics, ending up identifying a scientific discussion with any other discussion. Again, this link is too hasty. To cite another example, Galileo may have tried to convince Bellarmine with all kinds of rhetorical arguments, just as a politician tries to convince potential electors with all kind of propaganda, but Galileo's arguments are *proper* to the context of science because they depend on scientific dialectics. Take this basis away, and the hermeneutics of science immediately become a picklock for violently invading anyone's privacy.

The dialectical model provides us with an image of scientific practice that is perhaps less severe than that of the methodological, and less elastic than that of the counter-methodological model, but more realistic then both. We are no longer forced to choose between an algorithm and a chat at the bar, or, to change images, between "a dragon or a pussy cat."[31] The dialectical model of science promises something less comfortable than a cathedral but more dignified than a tavern on which to concentrate our cognitive efforts. We must now try it out.

5.2 Rhetoric and Relativism

Three questions in particular are crucial. In the dialectical model, the validity of an argument depends on factors that in most if not all cases change according to dialectical situations, while its strength depends on configurations of these factors, which are also variable. Consequently, a good argument in a situation in which a certain configuration holds will not normally be so in a different situation with a different configuration. The first question, then, is: will the dialectical model slip into the kind of relativism that claims every situation has its own criteria which cannot be compared?

The second question is related to the first. If in one situation a certain configuration of factors holds good, and, in another situation, another, how can one rationally choose between two theories in situations with different configurations? Won't the dialectical model end up concluding that the notion of rationality is useless and empty?

Then there is the third question. In the dialectical model, arguments aim to create, reinforce, and shift consensus over a certain theory or cognitive claim. One then wonders what the relationship between this consensus and the intrinsic value of that theory or claim is. Formulated in the most classical of terms, the question is: what relationship is there between rhetoric and truth?

The dialectical model appears to be caught between a rock and a hard place. If, against the evidence provided by the history of science, it insists

Table 5.1 Various Forms of Relativism

X	Y
(1) truth, rationality	society, culture[32]
(2) truth	tradition[33]
(3) standards	forms of life[34]
(4) opinions, standards	social context[35]
(5) meaning, truth and falsehood	style of reasoning[36]
(6) objects	conceptual scheme[37]
(7) reality, reason	language, way of social living[38]
(8) observations, meanings, standards	comprehensive theories, paradigms[39]

on claiming that there is a single configuration of dialectical factors, then it falls into the most dogmatic of methodological models. If, by contrast, it accepts what the history of science has to teach and denies there is a single configuration, then it seems to fall into the counter-methodological model.

Let us begin with relativism and ask: what (X) is relative to what (Y)? Table 5.1 contains a few answers taken from current literature on this subject.

This table is perhaps redundant and could be slimmed down.[40] What is clear, however, is that even in a slimmer version the list contains different positions. Although it is difficult to establish a hierarchy, there is here a scale which, read from top to bottom, starts with weaker and moves to stronger forms of relativism. In this scale, types (1) to (4), presumably very similar, are further up the scale than (5) because styles of reasoning can change even within the same society, tradition, or form of life. Type (6) is one step below the others but one step above (7) because it is possible that different conceptual schemes can live together within the same style of reasoning. Finally, type (8) is at the bottom of the ladder because the degree of variation of theories and paradigms seems to be greater than that of all the other entities in column Y. Some forms of relativism are more radical than others and some others are so weak as to be practically indistinguishable from certain kinds of objectivism (especially if one considers that objectivism also gives rise to a scale of positions). The whole discussion, however, is a little captious if the values of Y are not defined precisely or if they are interpreted in their intuitive meaning. It risks becoming circular, moreover, if the changes in Y are not defined independently of changes in X.

Relativism seems committed to maintaining three theses. The first two are:

(a) Every society, culture, or tradition has its own criteria (or its own world);

(b) There are no permanent, universal meta-criteria, common to different societies, cultures, and traditions.

The combination of (a) and (b), which we call the *contextuality of criteria* view, leads to precisely the conclusion that any theory is as good as any other because if theory T_1 is considered good by the criteria of a given society, culture, tradition Υ_1; and theory T_2 by the criteria of another society, etc. Υ_2; and if there are no common criteria to Υ_1 and Υ_2, then T_1 is as good as T_2.[41] Since this might look as though the choice between T_1 and T_2 is arbitrary, and the relativist does not like this idea, he replaces it with the following:

(c) The choice between two theories belonging to two different societies, cultures, traditions, etc. is not based on objective arguments and grounds.

In line with (c), the relativist substitutes the problem of the comparative appraisal of two theories with that of describing the transformation from one to another.[42] He maintains, for example, that we cannot say Galileo was more rational than Bellarmine or that he was any closer to the truth than his rival; we must simply take note that he introduced a new conceptual grid, and the question whether by doing so he was more or less rational or got closer to or further from the truth is simply "out of place."[43] I shall call this thesis the *redundancy of appraisal notions* view.[44]

The relativist seems to think of this view as a consequence of that of contextuality of criteria. Rorty, for example, starts off with (a) to get to (c):

> The world does not speak. Only we do. The world can, once we have programmed ourselves with a language, cause us to hold beliefs. But it cannot propose a language for us to speak. Only other human beings can do that. The realization that the world does not tell us what language games to play should not, however, lead us to say that a decision about which to play is arbitrary, nor to say that it is the expression of something deep within us. The moral is not that objective criteria for choice of vocabulary are to be replaced with subjective criteria, reason with will or feeling. It is rather that the notions of criteria and choice (including that of "arbitrary choice") are no longer the point when it comes to changes from one language game to another.[45]

Feyerabend, on the other hand, prefers to use (b) in order to get to (c). Thus, he writes:

> Critics of the idea that scientific debates are settled in an objective manner do not deny that there exist "means of deciding" between different theories. On the contrary, they point out that there are many such means; that they suggest different choices; that the resulting conflict is frequently resolved by power-play supported by popular preferences, not by argument; and that argument at any rate is accepted only if it is not just valid, but also plausible, i.e in agreement with non-argumentative assumptions and preferences.[46]

It is interesting to note that redundancy of criteria is more radical than the Cartesian dilemma. What it claims is that, since criteria common to different stages of Υ—that is, societies, cultures, traditions, etc. (theses [a] and [b])—do not exist, then an objective, argumentative contact between interlocutors belonging to these stages cannot exist, and consequently the normal notions of appraisal lose all meaning (thesis [c]). The opinion I intend to defend here is that the dialectical model of science allows us to shift between the horns of this dilemma. In the next section, examining the question of rationality, I shall return to this point more analytically with a historical case. In this section I shall be dealing with relativism.

Let us translate theses (a)–(c) into the dialectical model. They thus become:

(a′) Every dialectical situation has its own configuration of substantive factors;

(b′) A configuration of substantive factors common to all dialectical situations does not exist;

(c′) The choice between two theories belonging to two different dialectical situations is not based on objective arguments.

Although it is a historical matter to be examined case by case, we can easily accept both (a′) and (b′). If they hold, then as we have already seen, it may happen that an argument for theory T is strong in a dialectical situation s_1 in which the configuration c_1 is in force and weak in s_2 with a configuration c_2, and that consequently T is good according to c_1 and bad according to c_2. This is not relativism yet; it simply expresses the principle that the strength of an argument is relative to its premises, and that the quality of a theory is relative to the state of its discussion. Relativism crops up when the redundancy of appraisal notions view is introduced, and it is precisely this view that cannot be accepted.

Let us consider an abstract case. Suppose there is a controversy between two interlocutors A and B, upholders respectively of T_1 and T_2. Suppose, moreover, that A accepts configuration c_1 and B accepts c_2. For example, A may consider the agreement of a theory with known facts to be essential, while B may take the power of unification as more important. Or again, A may consider the derivability from a few simple principles most important, while B considers intertheory consistency more fundamental (this situation reflects a part of the debate between upholders of the steady-state and the big-bang theories). The point is: can A and B manage to agree on, say, whether T_1 is better than T_2?

It would seem not. In order for there to be an agreement, A has to produce better arguments than B, that is, he must defeat him in a debate. However, according to our explications in the previous chapter, the strength of an argument depends on its configuration, and, in our case, A and B do not accept the same one.

How will A and B behave, then? The counter-methodologist would say they will rely on tricks, propaganda, power-plays, etc. or else that everything depends on external factors. But there is no need for this. Given that one interlocutor has to defeat the other, it is enough for both of them to engage in a debate utilizing whatever argumentative means of persuasion will help achieve their aim. Clearly, if in such debate a sequence of the kind below is produced:

A	B
T_1 / r_1	$\neg T_1 / r_2$
...	T_2 / r_2
$\neg T_2 / r_1$	...

in which each interlocutor defends his own theory and rejects the other's on the basis of his own reasons, the debate stalls and neither interlocutor has any more hope of winning than he would if he shouted louder than his rival or pounded the table with his fists. A and B can try to get out of this situation, however, and they usually do. Since they are members of the same tradition and therefore share many of the same factors, if they have the will and desire to converse, they will begin to build an overlapping area between their respective configurations—a small house, a cabin even, where they can look each other in the eye and say "*we both.*" Thus they will set aside reasons r_1 and r_2 and look around for different ones. For example, they will cite other factors, adduce other values, invoke other assumptions, etc. Once the initial reasons have been taken away, the remaining configuration of substantive factors will either shrink or expand, and there is no reason to think that a debate of this kind cannot be held and brought to a conclusion. The moment one interlocutor, say B, loses the debate, the other theory automatically becomes the better one, supported by winning arguments. This confutes thesis (c′) because a choice between T_1 and T_2 conducted in this fashion is a choice made on the basis of objective arguments.

Why then does the counter-methodological model uphold thesis (c)? There seems to be two reasons.

First, because, under the influence of the Cartesian syndrome, the model concludes from the fact that universal criteria do not exist, that we have to become "critics of the idea that scientific debates are settled in an objective manner," to use Feyerabend's expression. This conclusion does not follow in the dialectical model, because to settle "in an objective manner" does not mean establishing impersonally (or from the perspective of God's eye) that one side is definitively right and the other irremediably wrong; it means, rather, establishing a dialogical contact between the two sides so that the shift in consensus takes place through a debate between them at the end of which one gives in to the other.

Second, because in the counter-methodological model theory-change is considered, holistically, as *inter*-traditional. The relativist assumes that the relationship between two scientific theories (for example, classical mechanics–quantum mechanics or theory of impetus–theory of inertia) is the same as that between two traditions (for example, western–Chinese medicine, scientific–Hopi culture). In the dialectical model, where the change is necessarily *infra*-traditional, this is not true. As I have already said, the factors of scientific dialectics define the scientific tradition or game, not this or that style of reasoning, conceptual scheme, system of thought. Thus they constitute a common framework for those who intend to stay within the tradition—which is, after all, the tradition of science from the ancient Greeks to our day.[47] It is true that the quality and order of these factors can change and indeed often do, but they are always of the same type. This means that interlocutors have the possibility of establishing a contact at any moment. In order to transform this possibility into reality, there is only one strategy: to look for a minimum configuration of factors from which a consensus can change through a debate.[48]

Popper has quite rightly pointed out that this shift depends on criticism, but in his praiseworthy efforts to fight relativism, he has put down as mythical the idea that "a rational and fruitful discussion is impossible unless the participants share a common framework of basic assumptions or at least, unless they have agreed on such a framework for the purpose of the discussion."[49] One of his arguments is that a comparison between two radically distinct languages reveals the limitations of each. Since this comparison can be made by using one of the two as a metalanguage, Popper concludes that "we are forced, by this comparative study, to transcend precisely those limitations which we are studying. And the interesting point is that we succeed in this. The means of transcending our language is criticism."[50]

This conclusion is acceptable but it does not prove Popper's thesis. On the contrary, if, as he says, criticism is the instrument we use to understand and overcome the limitations of our languages, then critical method is precisely (part of) that common framework that should be shared from the start if a discussion is to be possible. Thus, a framework must always be presupposed. Popper himself, when introducing his criteria of demarcation as "a proposal for an agreement or a convention" in his *Logic of Scientific Discovery,* wrote: "as to the suitability of any such convention opinions may differ; and a reasonable discussion of these questions is only possible between parties *having some purpose in common.*"[51]

The dialectical model tells us that, even if two rival sides have different aims, they still do and must share a set of factors that allows them to have a reasonable discussion.

5.3 Rhetoric and Rationality

The framework of factors of scientific dialectics is also useful for dealing with the question of rationality.

In the methodological model, rationality is defined in terms of abiding by certain rules. A typical explication runs as follows:

Expl. 1 A theory T is rationally acceptable if and only if it is $x, y,$ or $z,$

where $x, y,$ and z are properties such as "testable," "fruitful," "with high empirical content," and so on, specified by one methodological code or another, be it "Euclidean," "critical rationalist," "sophisticated," and so on.

An explication such as this clashes with what we have called the "paradox of scientific method," which shows it is possible to violate a rule without being irrational (in the sense that this violation leads at times to theories considered better than their rivals). A methodologist might then look for better, more "sophisticated" explications, but this inevitably leads to the failure of his enterprise: he is forced to chase his shadow, find ad hoc rules that cover such and such an episode after it has taken place. Moreover, there remains the problem that if one links rationality to rules, on the basis of what rules can a change of rules be considered rational? It looks as though the best thing to do is to sever that link.

The counter-methodological model has done precisely this, but by lopping off the branch on which it wanted to sit. Having denied that rationality can be defined in terms of respect for rules, it ultimately accepted the redundancy of appraisal. According to this model, being rational means accepting those opinions and theories that have prevailed *de facto,* for "reasons external to science" (Ronchi) or through "irrational means" (Feyerabend). Alternatively it means being "willing to pick up the jargon of the interlocutor" (Rorty). From this point of view, there is no point in wondering whether a theory change is rational or not; quite simply, one theory prevails, and it is rational to accept the theory that has prevailed.

But *how* does a theory prevail? Kuhn quite rightly introduced the idea of "conversion," but the concept still needs to be made more precise since several questions remain unanswered.

The first regards the *specific nature* of the conversion. Since conversions also exist in the domains of aesthetics, religion, ethics, politics, etc., how can scientific conversions be distinguished from others? It is not enough not to ask for "an epistemologically pregnant answer to the question 'What did Galileo do right that Aristotle did wrong?' any more than we should expect such an answer to the question 'What did Mirabeau do right that Louis XVI did wrong?'."[52] Even though there is no "epistemologically pregnant answer," in the sense that there is no impartial and impersonal method for providing it, the fact remains that the specific way in which Galileo defeated Aristotle is not the same as the way Mirabeau vanquished Louis XVI. It may be useful to draw resemblances, but to annul all differences only creates confusion.

The second question concerns the *quality* of the conversion. The word "conversion," in fact, can signify both the *final result* of a long process of thought (for example, the conversion of St. Augustine), in which case it is not unlikely that it was induced by an exchange of arguments, and a *psychological* form of a (sudden) change of heart as a result of some experience (for example, the conversion of St. Paul on the road to Damascus), in which case there is no exchange of arguments. To which of these two cases does a scientific conversion belong?

There is a third unresolved question. A scientist who undergoes a conversion does so for a variety of reasons. Lorentz was never converted to the relativistic interpretation of his equations, and Einstein was never converted to the "orthodox" interpretation of quantum mechanics. And yet by the early 1920s most theoretical physicists had been converted to Einstein's relativistic interpretation and were soon converted to Bohr's interpretation of quantum mechanics. Is there any way of saying that some conversions did not take place due to obstinate personal prejudices while others did? Or of saying that some conversions are good, critical, etc. while others are not? Here too, it is not enough to say that "scientific breakthroughs are not so much a matter of deciding which of various alternative hypotheses are true, but of finding the right jargon in which to frame hypotheses in the first place."[53] A reason must be found for why one kind of jargon is "right" and another wrong.

The dialectical model seems to offer such a reason. Unlike the counter-methodological model, it conserves a normative notion of rationality; but unlike the methodological model it links rationality not to certain *properties of theories* fixed by rules, but to the *quality of the arguments* which support the theories. The dialectical model replaces Expl.1 with the following:

Expl.2 A theory T is rationally acceptable if and only if it is supported by valid arguments, or if the arguments supporting T are stronger than those supporting T'.

This explication has several clear advantages. Compared to methodological rationality, dialectical rationality is ethically more *tolerant* because it is not linked to a single property or to a set of previously established requisites, but rather to a free debate over different properties and requisites. Compared to the rationality of "anything goes," dialectical rationality is more *adequate,* because it does not depend on the whims of authorities or on external social factors. This reflects actual scientific practice, where it may happen that a scientific community prefers theory T' to T even if T' does not explain more facts, anticipate "novel facts," solve more problems, etc. but where it never happens that it prefers T' if it is not supported by *stronger* arguments than those supporting T.[54] Compared to the rationality of the rule "respect good epistemic manners," dialectical rationality is prescriptively more *precise,* because there are objective factors (the substantive basis of scientific dialectics) that specify what these good manners are, and formal factors (the procedural basis of scientific dialectics) that specify how they should be respected. Finally, compared both to methodological rationality and what rationality remains in the counter-methodological model, dialectical rationality is philosophically more *attractive,* because it depends on its natural soil, that is, the strength of arguments: what does being rational mean if not following the best argument?

The controversy over Copernicus' theory illustrates the virtues of Expl.2. With regard to this controversy, in particular that aspect of it regarding the reliability of telescopic observations, Ronchi has written: "Galileo knew, and he declared as much in no uncertain terms, that there was no scientific demonstration able to prove him right or wrong; it was simply a matter of faith. He solemnly and tenaciously reasserted that faith in observation, both direct and with instruments, which had been condemned to death by Platonism two thousand years before. This is why he set out on a propaganda campaign as if he was propagating faith, and he never engaged in a scientific or technical discourse."[55]

Referring to another aspect of the same controversy, Feyerabend has similarly claimed that Galileo had no incontrovertible proof at hand: "How does he proceed? How does he manage to introduce absurd and counter-inductive assertions, such as the assertion that the earth moves, and yet get them a just and attentive hearing? . . . For Galileo uses *propaganda*. He uses *psychological tricks* in addition to whatever intellectual reasons he has to offer."[56]

Feyerabend still speaks of "intellectual reasons." Rorty goes even further:

> Galileo, so to speak, won the argument, and we all stand on the ground of the "grid" of relevance and irrelevance which "modern philosophy" developed as a consequence of that victory. But what could show that the Bellarmine-Galileo issue "differs in kind" from the issue between, say, Kerensky and Lenin, or that between the Royal Academy (c.1910) and Bloomsbury? . . . At this point, it seems to me, we would do well to abandon the notion of certain values ("rationality," "disinterestedness") floating free of the educational and institutional patterns of the day. We can just say that Galileo was *creating* the notion of "scientific values" as he went along, that it was a splendid thing that he did so, and that the question of whether he was "rational" is out of place.[57]

When pain is acute, normal perception is dimmed. The symptoms of the Cartesian syndrome are clearly so piercing here that our authors cannot perceive that texts and history tell a different story. Since this is not the place to deal with the whole Copernican controversy, I shall limit myself to a partial but nonetheless essential aspect of it: the question of sunspots.[58] My aim is to show that even when, as a result of profound changes in certain basic assumptions, methodological rules are incapable of settling a controversy (either because they are not shared, or because they are interpreted differently), the basis of scientific dialectics provides common ground on which a rational, critical shift based on "good reasons" can come about.

When in March of 1611 Christopher Scheiner directed an "optic tube" towards the Sun, he chanced upon "new and almost incredible phenomena"[59] which gave him "first enormous admiration and then pleasure too": on the Sun, he noted, there were "every now and then certain spots resembling black drops." Struck by this, he continued to observe, rejecting several hypotheses before advancing two theses: the spots are not on the Sun, and the spots are stars. Galileo, independently and contemporaneously observed the same phenomena and arrived at opposite conclusions: the spots are contiguous to the body of the Sun, and are clouds.

Both sets of conclusions are dictated by the respective commitments and intellectual orientations of the two scientists. Not only is the Ptolemaic theory jeopardized (if the sunspots rotate around the Sun, then not all the heavenly bodies rotate around the Earth), but above all the basic assumption of Ptolemy's theory—the incorruptibility of the heavens—is at risk: if the spots change shape and size, how can one exclude alterations in the heavens? Scheiner is explicit on this point when he writes that it had "always appeared unseemly and highly unlikely that the spots lie on the surface of the Sun which is a very shiny body," and when he claims that his aim is to "free the Sun of the offensive spots."[60] "Who would place clouds on the Sun?" he asks again.[61] Although the question may seem ridiculous today, it is no more so than the question, "Who would place positive electrons around an atomic nucleus?" before the discovery of antiparticles. As W. Shea has written, for Scheiner "the spots *had to be* stars."[62] Likewise, one could say that for Galileo, the spots *had to be* clouds or alterable bodies. Different assumptions lie behind these rival positions. Scheiner's assumption starts at the core of the Aristotelian-Ptolemaic system: the heavens are incorruptible. In order for his own sunspot theory, and consequently Copernicus' heliocentric theory, to be accepted, Galileo had to knock this assumption down.

Shea has also spoken of the "bewitchment of theory"[63] that tacitly induces people to view phenomena according to their expectations. However powerful this effect may be, it has its limitations, as the development of the controversy shows. Galileo uses, among others, two kinds of argument against Scheiner. The first can be found at the end of his *Second Letter* (August 14, 1612) to Scheiner. It is an argument by retort. On what is his opponent's assumption of the inalterability of the heavens based? On experience. Thus, if experience shows that the heavens are indeed alterable, why not accept the conclusion? Galileo writes:

> Now, in order to gather some fruit from the unforeseen marvels that until our time have been hidden from us, it would be a good idea for the future if we go back to lending an ear to those wise philosophers who regarding the celestial substance were of a different opinion to Aristotle and whom Aristotle himself would not have shirked if he had the possibility to make similar sensory observations. For he did not simply include manifest experiences among the most powerful means for drawing conclusions about natural matters; he gave them pride of place [. . .] Rather, I go still further, I estimate that I am less against Aristotle's doctrine than those who would like to maintain the incorruptibility of the heavens in any case, because I am sure that he never took the proof of their inalterability, more certain than this, that any discourse should bow to evident experience.[64]

Galileo astutely stretches the situation to fit his needs, because his "sensory observations" or "evident experiences" of sunspots are far from clear and it is dubious whether Aristotle gave experience "pride of place."[65] He does not however stretch the point to the extent of making his argument useless; he cites at least one source that belongs to his rival's tradition and implicitly invokes the presumption of reliability—if a procedure has been successful in the past, then it is reliable. Thus Scheiner's argument from authority (Aristotle) is challenged by Galileo's counter-argument from experience, the authority of the authority.

The second type of argument that Galileo uses is *ad hominem* in the classical sense and is found towards the end of the *Third Letter* (December 1, 1612) to his adversary. Here Galileo writes: "Now Peripatetics, in order to protect themselves against the imminent threat of the alterability of the heavens, leap to the defence by saying that sunspots are stars; in the meantime they leave a thousand other entrances open to enemy assault, because the number of planets being seven, their rotation around the Earth, the regularity of their movements etc. are no longer defended."[66]

Here is what I shall call a "dragging strategy" at work. The list below, contrasting the arguments of Scheiner and Galileo, sets out the relevant moves. Let I = the assumption of the incorruptibility of the heavens, S = sunspots are stars; T_1 = there are only seven planets; T_2 = planets rotate around the Earth; T_3 = the planets are very regular. Thus, Galileo points out that the sets $\{I,S\}$ and $\{T_1,T_2,T_3\}$ are incompatible. At this point, the opponent stands at a crossroad: either he withdraws T_1-T_3 conserving S, or he withdraws S in favor of Galileo's explanation of the spots. In the second case, he is forced to withdraw I as well, because if the spots are clouds, then the heavens are not incorruptible. In the first case he can maintain I but exposes himself to a new attack. Withdrawing T_1-T_3 is tantamount to knocking down most, if not all, of the Ptolemaic edifice. At this point what will I lean on? The dragging effect of the strategy consists precisely in this: by knocking down a theory, the assumption on which that theory is based is "dragged" down with it.

Scheiner	Galileo
$I \wedge S \wedge T_1 \wedge T_2 \wedge T_3$	$(I \wedge S) \rightarrow \neg(T_1 \wedge T_2 \wedge T_3)$
$-T_1$	—
$-T_2$	—
$-T_3$	—
	$(I \wedge S)$?

Does this strategy clinch the issue? Not at all, because assumption I does not logically imply theses T_1–T_3, and Scheiner could withdraw the latter without feeling obliged to renounce the former. Nevertheless, the blow it delivers is a forceful one. Theses T_1–T_3 give support to *I*, and if such support is missing, the whole construction teeters and ultimately collapses: how can one go on defending the assumption of the incorruptibility of the heavens if the very theses this assumption is connected with fall down?[67]

We can now pick up the question of rationality where we left off. It is true that Galileo "converted" a number of people, but this conversion, if we really must use the term, was not the result of tricks, propaganda, or because Galileo "just lucked out." Our analysis shows that these conversions came about because Galileo put forward strong arguments. It is also true that Galileo did not succeed in converting everyone; Scheiner felt the blows but never renounced his view. Our analysis suggests that after his belligerent exchange with Galileo, Scheiner's resistance had mainly (though not only) to do with personal obstinacy. In terms of arguments, he had very little room for maneuver because once he had withdrawn theses T_1-T_3, his main theory became less credible.[68] It is true, finally, that Galileo introduced a new conceptual "grid." Here again our analysis shows that the question of whether he was rational or not is *not* "out of place." Although there is no supreme court of reason before which to bring Galileo's and Scheiner's grids, there are at least certain factors of the scientific dialectics common to both and in the light of which one position appears stronger than the other.

Once again, the error lies in thinking that if there is no universal method then we can only accept whatever people do and try to understand their jargon. This is the same mistake as claiming that since all legal systems are fallible, justice should be taken into individual hands. At least in science, doing justice means using good arguments.

5.4 Rhetoric, Truth, Assertibility

The last thing we must deal with in this chapter is delicate. Rhetoric has a biblical (if Plato is a Bible for philosophers) condemnation hanging over its head: what has it to do with truth? Does the fact that an interlocutor, a group, a whole community, is persuaded to accept a thesis *T* have anything to do with *T* being true?

I have already recalled in the first section of this chapter how the opposition between persuasion and truth was used by the founding fathers of modern science as a weapon against dialectics. Shared by rationalists and empiricists alike, such opposition appears to be a cornerstone of Western philosophy. It can be found in Hobbes, Locke, Kant, Schopenhauer, and in our day in Perelman—the very philosopher who has contributed most to calling our attention to rhetorical argumentation.[69]

Even those who agree in principle that rhetoric has a role to play in science have made sure that this role does not concern the question of truth. For example, it has been claimed that justification (in the sense of persuasion, as distinct from both "demonstration" and "verification") "regards neither inference nor truth, but the validity of the rules and procedures we follow,"[70] or that science requires justification of its "activity as a whole," as well as of its "criteria, methods, rules, procedures, axioms," but not of the truth of its specific cognitive claims.[71] As we have seen, this was also Perelman's general position, although he came to admit that persuasive argumentation comes into science in different contexts, for example, when new hypotheses are formulated,[72] when one has to decide whether a given "unsatisfactory" theory can be introduced into the scientific domain,[73] and when "new facts must be integrated with a theory."[74] And this seems to be the position of those who consider rhetoric instrumental to scientific truth.[75] But in general it is maintained—even before Kant and after Quine—that truth and falsehood do not depend on the consensus of the audience but on the "tribunal of experience,"[76] or to put it differently, that "the experimental sector is the only place the absolute criterion of truth is based on, and is both the testing ground and the last court of appeal for rival theories."[77]

Although this view prevails among scientists and philosophers, especially those who hold "realism about theories,"[78] and although it looks quite reasonable, it is nevertheless disputable. If we believe or maintain that at least some of our scientific claims are true, then, since our acceptance of these claims as part of the body of scientific knowledge depends on a debate along the lines of the dialectical model, truth is to be connected with rhetoric in some way or another. This connection, however, cannot be straightforward. After all, Plato was right when, in his polemics against rhetoric, he argued that people can be convinced with rhetorical arguments, or, as he wrote, by exercising an appropriate influence on the "psychology of the masses"[79] to the effect that a donkey is a horse, even though a donkey is not a horse.

A distinction is then needed. A claim such as "*T* is true" raises two questions: on the one hand, the question of judging that *T* is true, which concerns the *ascription* of truth we make; on the other hand, the question of *T* being true, which concerns the *definition* of truth we adopt. The former is the question "How do we know it?"; the latter is the question "What does it mean?"

Take definition first. Traditionally (and typically among realists), "true" is considered a predicate that refers to a substantive property allegedly possessed by all true assertions. The correspondence theory of truth, which is the oldest and most venerable of theories that treat truth in this way, offers the following definition:

(C) *T* is true if and only if it corresponds to the facts.

Since the difficulties this definition encounters are well known, I do not need to go into any detail here. All I need to do is remember that the correspondence relation is elusive because it is not clear what is meant by an assertion (which is a mental act, a sign, or a sound) "corresponding to" a fact (which is a slice of reality); and that the correspondence relation is not univocal, because the same fact can be described by, and put in correspondence with, incompatible assertions. The result is that the correspondence theory does not provide an intelligible definition of truth. Not even from the viewpoint of God's eye can we understand what correspondence between facts and assertions amounts to.

Things are no better if we take "true" as a predicate referring to an epistemic property. The theory of truth as ideal rational justification[80] offers the following definition:

(J) *T* is true if and only if it is justified in ideal conditions.

Here again, the difficulties are well known. To name but a few: the phenomenon of the underdetermination of theories also holds in ideal conditions because two equivalent but incompatible theories can equally well be justified;[81] ideal conditions cannot be considered the limit of a series of actual conditions because there is no guarantee that such a series has a limit; we cannot extrapolate from what is the case in actual conditions to what would be the case in ideal conditions, because there is no principle of induction to bridge the gap, and because, in any case, applying a principle of this kind implies knowing that what is the case in actual conditions is *true*.[82] The result here is that the ideal rational justification theory of truth does not provide a correct definition of truth. Not even from the viewpoint of the ideal community could we say that a certain claim that is justified is true.

These negative results of the two main substantialistic theories of truth suggest that truth should not be taken as a substantive property (either epistemic or not). As a consequence, persuasion should not enter into the definition of truth. In this sense, there is no relationship between rhetoric and truth.

Things change if we turn to our ascriptions of truth. Here rhetoric clearly has a constitutive role and persuasion is essential because our ascription of truth to *T* depends on the outcome of a debate conducted according to the factors of scientific dialectics. Since, as we know, facts are among these factors and are not the only ones, experience cannot be taken as *the* tribunal of (our ascriptions of) truth, but as one of the witnesses that are cross-examined by the participants in the debate. In the dialectical model of science there is no tribunal apart from discussion.

As an essential ingredient of our ascriptions of truth, persuasion works as an *indicator* of truth. Realists themselves agree on this. According to Popper, for example, since we can never "produce valid reasons (positive reasons) in favor of the truth of a theory,"[83] the degree of corroboration of a theory (which is a degree of persuasion when translated into the terms of our game with three players) cannot be taken as a measure of its verisimilitude but as "an indication of how its verisimilitude *appears* at the time *t*, compared with another theory."[84] Likewise, according to Putnam, since "there is no single general rule or universal method for knowing what [actual] conditions are better or worse for justifying an arbitrary empirical judgment,"[85] the degree of justification of a theory at the time *t* (again a degree of persuasion) cannot be considered but as an indicator of how the ideal justification appears at *t* to a certain community.

We can now perhaps be more precise about the relationship between persuasion and truth. Suppose *T* is rationally acceptable in the sense provided by Explication 2, that is, in the sense that *T* is supported by valid arguments or by arguments that are stronger than those against *T* according to factors holding in the dialectical situation in which the debate about *T* takes place. Then *T* is persuasive for the community. But if persuasion is an indicator of truth, we can take rational acceptability (or assertibility) as a (sufficient) condition of *T*'s truth according to this schema:

If *T* is rationally acceptable then *T* is true.

Note that "*T* is rationally acceptable" does not mean "*T* is true." The schema above does not define truth, it only licenses our ascriptions of truth and, by so doing, provides a *presumption* of truth.[86] As to the definition of truth, once we reject substantialistic theories, we have two alternatives. The first is eliminating the word "true" from our vocabulary altogether, considering it an "empty compliment" with no other use but as an "expression of commendation."[87] This move is unacceptable because the concept of truth plays an essential role in deductive logic, in ordinary language (without it certain propositions such as "Everything my wife says is true" could not be expressed), and even in science. The second alternative consists in making the problem of the definition of truth less dramatic. The disquotationalist or deflationist theory of truth serves this purpose well.[88] It offers the following definition:

(D) "*T*" is true if and only if *T*.

Apart from its simplicity, this theory recommends itself for two main reasons. In the first place, it saves the fundamental intuition that truth does not change over time. According to (D), truth is not a substantive property as in (C) and (J) which people may believe they have grasped at time *t* and about which they may change their minds at a later time *t'*: "true" is any claim that satisfies (D). Therefore, if *T* is the case, or *T* is asserted, *T* is true independently of time. Second, the disquotationalist theory does justice to the realist's intuition about the aim of science.

The realist is committed to the view that science aims at true theories. J. Watkins, for example, has written that "to say that truth is no part of the aim of science is on a par with saying that curing is no part of the aim of medicine or that profit is no part of the aim of commerce."[89] The disquotationalist theory is not incompatible with this view; actually it allows it in a harmless way. Take medicine. Should one say its aim is health, with the risk of getting involved in the difficult if not impossible task of defining what health is? To all theoretical and practical purposes, it is enough to say that medicine aims at knowing whether high cholesterol causes heart disease, whether the HIV virus leads to AIDS, whether genes are responsible for schizophrenia, and so on. Similarly, take science. Should one say it aims at truth, with the risk of looking for a certain, ineffable substantive property which all true propositions are supposed to have in common? It is enough to say that science aims at knowing whether snow is white, whether planets rotate in elliptical orbits, whether bodies attract each other, and so on. Once such claims are asserted, we can say that they are true in the way allowed by (D). With this move much is gained and nothing is lost; in particular, the realist's intuition of an independent world is not lost. Rejecting substantialistic theories of truth does *not* put us in the hands of antirealism or idealism, but still leaves us on realist territory. This raises a question, though, that requires a longer discourse to be deferred to the next chapter.

Epistemology and Rhetorical Strategies

The Manager. Let's come to the point. This is only discussion.
The Father. Very good, sir! But a fact is like a sack; when it is empty it won't stand up. If you want it to stand, you have to fill it with the reason and sentiment that gave it life.

L. Pirandello, *Six Characters in Search of an Author*

Everything that is sensible exists as it were in several ways. Everything that is real is bound to an infinity of sequences, fulfills a thousand functions; it carries with it many more characteristics and consequences than an act of thought can embrace. But in certain cases, and for a certain time, man subjugates this manifold reality, and triumphs over it to some extent.

P. Valéry, *Eupalinos, or the Architect*

6.1 Theories, Assumptions, Categories

In the dialectical model, theories and facts are among the substantive factors of scientific dialectics. Nevertheless, the model in itself is no guarantee that there will always be a stock of facts readily available for interlocutors to use in settling their differences.

As regards this question, the methodological model is often associated with an empiricist epistemology stating that experience is more or less pure data, independent of, or separable from, the theories being tested. For this reason, data function as a tribunal, an "impartial arbiter of scientific controversy."[1] The counter-methodological model, instead, is associated with an opposite epistemology I prefer to call hyper-criticist, due to its excessive use of Kant's idea that theories and facts cannot be separated.[2] On the one hand, theories are "all-pervasive," "ways of looking at the world," to the extent that "their adoption affects our general beliefs and expectations, and thereby also our experiences."[3] On the other, facts are so loaded with theories that "*hypotheses facta fingunt.*"[4] The most radical, but most coherent, conclusion of this epistemology is that "each theory will possess its own experience, and there will be no overlap between these experiences."[5]

If this conclusion were unavoidable, our scientific dialectics would be no more than the art of verbal persuasion and would deserve the same harsh criticism once reserved for the old dialectics. We must therefore make sure that the voice of experience does not disappear, reducing the game with three players to a monologue or, at the most, to a dialogue between interlocutors free to say whatever they believe about nature. To this aim, we must provide an appropriate epistemology for our dialectical model.

This will be the task of the present chapter. My first goal will be to establish the kinds of theories and the status of facts. My second will be to examine the role played by facts in scientific debates. This will lead to the discovery that in most cases of theory change it is possible to settle a controversy between supporters of rival theories with shared facts; that in some cases facts cannot decide an outcome; and that in others facts cannot be used at all. I shall then try to show how, especially in the latter cases, a rhetorical discussion based on other factors of scientific dialectics is the only way to express a rational theory-preference.

Let us begin with theories. To consider all of them as mere "ways of looking at the world" is misleading because there are, in fact, three types of theories at three different levels. As two have already been examined in chapter 4, I shall provide only a brief reminder.

The first level is that of *explanatory theories* (E-theories). As we have seen, these are the usual hypotheses formulated in theoretical terms that explain facts and regularities and that, once consolidated, become part of the stock of accepted available knowledge.

The second level is that of *interpretative theories* (I-theories). We took these as assumptions, that is, ontological interpretations of the world and their domains and we classified them as either general or disciplinary. As we have seen, the term "assumption" is intended to cover several methodological entities such as "paradigms," hard cores of "research programs," and so on.[6]

At the third level lie what may be called *categorial theories* (C-theories). In Kantian terminology, these are forms of perception or categories of the intellect; but the sense in which I take them here is different from Kant's.[7] I consider C-theories from an evolutionary point of view, and I conceive them as conceptual ways of adapting to the world incorporated in the physiology of our perceptive organs.[8] Phylogenetically, they are acquired, but ontogenetically they are a priori. Their function is different from that of other types of theories. C-theories do not dictate the basic entities of the world (ontology) as I-theories do, nor do they explain their behavior as E-theories do. Rather, C-theories provide the laws or basic rules according to which a set of phenomena or data can become an object or event in our cognitive environment. Our C-theories are therefore stable over time: alternative categories are hard to imagine; they could barely be considered human, still less those of our civilization.

C-theories presumably include the forms of perceptual space and time, the distinction between figure and background, the completing of figures (for example the "tendency towards pregnancy"), as well as certain conceptual schemes such as causality, substance, etc. Establishing their genesis, areas of application, and list is a complicated matter belonging to the sphere of empirical research.[9]

From a philosophical point of view, however, it is useful to stress a point: C-theories do *not* pose limits on E-theories and are not affected by them. On the one hand, it is possible to construct successful explanatory theories that do not contain the usual euclidean physical space or that do without the classical concepts of cause and substance. On the other hand, centuries of Copernican education have not altered the fact that we see the sun rising on the horizon at dawn, that the theory of relativity has not changed our visual space, that quantum mechanics has not abolished descriptions in terms of cause and substance from our everyday experience.[10]

Although this by no means proves that the system of categories is unique and fixed, as Kant understood it, there are some important consequences for the problem of theory change.

The first consequence is that every E-theory is associated with at least one C-theory and one I-theory. Take, for example, Aristotle's explanation of falling bodies. First, an object—such as a stone moving in a certain way in a certain direction—is taken as given. This requires a certain spatial C-scheme. Next, this object is attributed to a class—in Aristotle's version the class of falling bodies—which requires a certain I-interpretation concerning the directions of the universe. Finally, a reason for the behavior of these bodies is provided—for Aristotle that they accelerate the closer they get to their natural place—and this is an E-theory.

The same is true for other explanations. For example, Descartes' (E-) theory of vortex is associated with the conception (I-theory) that the world is a mechanism of bodies in contact; Newton's (E-) theory of gravitation is associated with the view (I-theory) that the world is populated by forces acting at a distance; Darwin's (E-) theory of natural selection is associated with the idea (I-theory) that in nature there are no transcendent causes at work; and so on. Therefore, when theories are said to be "all-pervasive," this applies only to C-theories; and when they are said to be "ways of looking at the world," this applies only to I-theories. But E-theories are neither one nor the other.

The second consequence is that only E-theories can properly be said to be empirically testable. Here there are some asymmetries. E-theories can be either confirmed or falsified by experience because they provide explanations. I-theories, on the other hand, being ontological conceptions, can neither be confirmed nor falsified by experience (at least not directly—

remember Galileo attempting to "falsify" the Ptolemaic-Aristotelian assumption of the incorruptibility of the heavens). C-theories, finally, can only be confirmed, because they are conditions of our experience. It follows that if a preference between two E-theories associated with two different I-theories is to be expressed, facts can only be a part of the argumentation.

The third and final consequence is that if we accept a stratified view of theories, we are also committed to a disparity view of theory change. Here there is another asymmetry. Changes at I-level bring about at least some changes at E-level, but not vice versa. It is a historical, contingent matter that a change at E-level is accompanied by changes in the higher strata.

We must now look into what this stratified view of theories implies for the status of facts.

6.2 Objects and Facts: Reference without Correspondence

The aim of E-theories is to inform us about the world. If a theory is well confirmed by observational and experimental evidence, then, generally speaking, we feel entitled to believe it tells us what reality is like. Considering things in a less general light, however, this view creates ambiguity and becomes a source of frustrating expectations because observation and experimentation prove to be obstacle courses to reality. If "telling us what reality is like" means, as the realist would like it to mean, "depicting" or "representing" or "providing descriptions that *correspond* to reality," then observation and experimentation are also dead ends. Nevertheless, they are all we have and are by no means useless; even if they do not give direct access to, or correspondence with, reality, they still lead us to a solid world of entities (objects and facts).

Let us take observation first. Consider an individual I (say, Galileo) claiming he has observed sunspots. This claim requires I to see certain phenomena, $a_1, \ldots, a_n$ (e.g., a luminous disk, areas of shadow, etc.), and to describe these phenomena with the predicate A = "sunspot." The structure of the observational process can be set down as follows:

(I) I sees $a_1, \ldots, a_n$;
I sees that $a, \ldots, a_n$ are A.

In this process there are two types of seeing, usually called, respectively, "nonepistemic seeing" and "epistemic seeing," or simply "seeing" and "seeing that."[11] If it is not considered a mere physiological condition (in which case it does not even deserve to be qualified as "seeing"), nonepistemic seeing depends on C-theories. As Kant held, our experience (or perception) of a_1 could not be distinguished from our experience (or perception) of a_2 if both were not placed within a spatial grid. Nor could a_1 be considered as coming before or after a_2 if the perceptive field were not organized along a temporal grid. Thus, "seeing" is at least C-dependent.

On the other hand, epistemic seeing requires the objects of perception $a_1, \ldots, a_n$, to be covered by a concept (sunspots) which in its turn requires theories. A person who knows nothing about astronomy cannot see that the sun has spots because he does not possess a concept that refers to his perceptions. Since providing concepts is the function of I- and E-theories, "seeing that" is I- and E-dependent. All the more so when one analyzes more complex observational descriptions such as "an electron leaps out of its orbit."[12]

The conclusion of this analysis is that an observational report such as "The sun has spots" is necessarily theoretical because, in addition to certain perceptual data, it contains conceptual elements. If it is theoretical, though, how can we say that it describes or depicts reality? The report "The sun has spots" *talks about* reality, about something that does not depend on us. This is proved by the fact that the concept it adds to perceptual data is not completely free, and anyone who wanted to change it would face serious obstacles which, beyond a certain point, would become insuperable resistance in a debate.[13] For example, I can see that $a_1, \ldots, a_n$, in addition to A_1 = "sun with spots," are perhaps A_2 = "sun perturbed by the atmosphere," but I cannot see that they are, say, A_3 = "dark butterfly in the sky." And yet, "talking about reality" is not the same as "describing (or providing an assertion corresponding to) reality," that is, grasping reality as it is in itself, because it is possible for mutually exclusive concepts to be compatible with the same perceptual material. It is true that without $a_1, \ldots, a_n$ there can be no A, but it is not $a_1, \ldots, a_n$ that determines *which* predicate (A_1, A_2, etc.) is most fitting. The relationship between $a_1, \ldots, a_n$ and each of these predicates is material, causal, not conceptual. A holds as long as the theoretical knowledge on which the seeing and seeing-that with which it is introduced holds.

With experimentation, things are no different. Imagine an experimental situation in which an individual I accomplishes a set of operations, $o_1, \ldots, o_n$ (for example, takes a glass jar, rubs it with a silk cloth, touches it with the tip of his finger, etc.), achieves a set of results $r_1, \ldots, r_n$ (for example, obtains a spark, feels an electric shock, etc.), and finally reports having seen B (for example "the jar is electrically charged"). The process is as follows:

(II) I accomplishes $o_1, \ldots, o_n$ and obtains $r_1, \ldots, r_n$;
I sees that $r_1, \ldots, r_n$ are B.

Here too the process has two stages, and both are theory-dependent. The first because accomplishing $o_1, \ldots, o_n$ and recognizing $r_1, \ldots, r_n$ as results presupposes, apart from the causal conceptual scheme, some E- or I-theories or expectations concerning the type of operations and results in question. The second because seeing that these results are B presupposes some knowledge about the predicate. *B talks about* reality because it is causally determined by the experimental situation $o_1, \ldots, o_n - r_1, \ldots, r_n$, but it cannot be said to describe reality because B is not conceptually determined by the experimental situation.[14] B holds as long as the theoretical knowledge on which the whole process depends holds.

The difference between *talking about reality* and *describing reality* can be illustrated in terms of reference. Let us introduce the following:

Criterion of reference: a term refers to an entity if and only if it gives an account of a public, observable, or repeatable experimental situation.

Since any observational or experimental account relies on at least one theory, by this criterion reference is ensured by a quintuple such as:

$$< S, O, R, P, T > = A$$

where S is an instrument, O a set of operations, R a set of results, P an element of the Cartesian product $O \times S \times R$ (indicating that a certain operation O has been carried out with a certain instrument S and a certain result R), and T an E-theory.[15] On the basis of this criterion, *that* A refers to something depends on $<S, O, R, P>$, but to *what* it refers depends on $<S, O, R, P, T>$, in particular on T. Let us call the reference depending on T *putative* or *theoretical reference* and the one that does not depend on T *real reference*. If A has putative reference, A "talks about reality"; if it also has real reference, then A "describes (or corresponds to) reality."

Putative reference is an element of reality as established *through* certain experimental operations and *within* a given theoretical framework that interprets the results of these operations. Real reference, on the other hand, is an element of reality in itself: it is what it is and exists or not regardless of the state of our knowledge. While putative reference always has an index (for example, A_1 = "Rutherford's electron," A_2 = "Bohr's electron," etc.), real reference has none. Hence, real reference is inaccessible, inscrutable. Nor can we hope to grasp it by omitting T from the quintuple and relying on observational and experimental situations alone, because if T is omitted, A has no reference at all,[16] while if T is introduced, A has only putative reference. Nor can we base ourselves on the success of a given putative reference in order to proceed to real reference: this would be a *metábasis eis allo ghénos,* a shift from the genre of concepts to that of reality in itself.[17]

In short, we can only know putative reference, that is, reference without correspondence.[18]

Must we then abandon the intuition that science allows us to know reality? This is hard to admit, because independent reality is the intentional target of all science's efforts. Or must we give up the intuition of independent reality altogether? This is also hard to concede, because without reality science cannot be conceived. The only thing to do, as with truth, is to play down the problem of reality. We can leave real reference to its destiny and apply, again, the deflationistic instrument, according to the scheme:

"*A*" refers to *A*.

In this way, both intuitions are safe. We can maintain the concept of independent reality and continue to say that science aims at knowing it, specifying, however, that what this really means is that science aims at knowing such things as electrons, genes, etc. Once we know that electrons, etc. exist, there is nothing else to say about the claim, "electrons are real"; just as once we know that snow is white, there is nothing to say about the claim, "'snow is white' is true." What there is left to add has nothing to do with reality in itself, or real reference, but with other problems. In particular, the following three: When can we say that we know that electrons, genes, etc. exist? What happens to these entities when theories change? How do theory changes take place?

These are the problems I will treat in the next three sections.

6.3 Objects and Facts: Empirical Realism

The idea that real reference is inscrutable may have a drawback but it also has an advantage: while we lose (correspondence with) reality in itself, we acquire (knowledge of) *objects* and *facts*.

I shall illustrate this point with a few examples, starting with observations. Let "The sun has spots" be the private observational report of an individual *I*. Once he has made it, *I* brings this report to the linguistic market of experts in physics and astronomy where he encounters both buyers and critics. A debate ensues. If the debate favors him—for example, if, after an exchange of arguments and counter-arguments, everybody answers yes or reacts in the same way to the question, "Are there spots on the sun?" or to the command, "Look at the spots on the sun!"—then one can say that sunspots are an object and that the sun has spots is a fact.

Let us take an even simpler case and see how, in everyday life, a predicate as subjective as color can become objective. An individual *I* sees something, identifies it as "green," and reports, say, *A* = "This lawn is green." So far "green" is not objective and *A* is not expressing a fact. *I* then takes

A and puts it on the linguistic market of ordinary people, ready to pin himself down in a debate. If in this debate *I* obtains the consensus of his interlocutors—for example, if at the command "Do not tread on the green lawn!" they all react in the same way; if to the question "What is the color of this lawn equal to?" they all point to the first stripe of the Italian flag; if to the assertion "the color of this lawn is darker than that of this carpet" they all nod in agreement—one can say that "green" becomes an object (an objective property) and that *A* expresses a fact.

The same is true for experimentation. Take Franklin's famous experiment as a first example. Two individuals *X* and *Y* are isolated by being placed on a wax support. The following results are obtained. r_1: when *X* rubs a glass jar and *Y* touches it, both individuals are electrified; r_2: when *X* and *Y* are touched by somebody who is earthed while he rubs the jar, neither are electrified; r_3: when *X* and *Y* touch each other after one of them has drawn a spark, both feel a shock; r_4: after this shock, neither of them are any longer electrified. Now, if everyone can repeat the same operations with the same results r_1–r_4, and if these results are interpreted theoretically in the light of Franklin's theory (according to which *X* passes electricity to the glass jar and, being isolated, receives none from the earth, while *Y* receives electricity and, being isolated, holds it), then one can say—in Franklin's words—that "electrical fire was not created by friction but collected";[19] that is—in our terms—that the properties of positive and negative electricity are objective.

Another example that will prove useful later is Newton's well-known experiment with a prism. A white ray of light penetrates a dark room through a small crack in a window and passes through a triangular prism of glass. The following results are obtained. r_1: on the screen placed beyond the prism, a colored oblong spectrum is formed; r_2: the colors of the spectrum are always in the same order from violet to red; r_3: when the size of the crack and the thickness of the prism are changed, the shape of the spectrum and the order of colors are unchanged; r_4: when the colored rays are sent through a second prism placed after the first but upside down, a spot of white light on the screen is obtained; r_5: when a single color is isolated by making a small hole in the screen and then making the light go through a second prism, a further refraction is obtained with no change in color. Now, if all these experiments are repeatable with the same results r_1–r_5, and if one accepts Newton's interpretation that rays are primary qualities, then one can say that the statement, "colors are simple rays with different angles of refraction," is a statement of an objective fact.

The conclusion to which these examples lead can be encapsulated in two definitions. The first is: *an object is the putative reference of a concept about which there is consensus.* The second is: *a fact is a shared state of objects.* For example, electricity is that particular "thing" to which the corresponding concept incorporated in an accepted theory refers; and the fact that electricity attracts little pieces of straw is the behavior of that "thing" that can be derived from, or explained, in terms of such a theory. Since the "thing" is that which results from a material (observational data, results of operations) perceived in terms of C-theories and conceptualized in terms of I- and E-theories, objects and facts are theoretical constructs. Thus, the object sunspots and the fact that the sun has spots are theoretical constructs based on, or starting from, visual perceptions; likewise, the object electricity and the fact that the jar is electrified are theoretical constructs based on, or starting off with, material operations; and so on.[20]

Objects and facts cannot be taken separately, in the sense that, for example, first objects then facts are construed. The construction of an object goes hand in hand with that of a fact, because an object is the putative reference of a concept and, as Kant pointed out, a concept is a predicate of a possible judgment. Thus, the object electricity can be construed in physics when its concept figures as a predicate in a judgment such as "the glass jar is electrified" which in its turn construes a fact out of accepted observations and interpretations.

Conceived in this way, objects and facts guarantee that science is as objective as it is able to be. Science is objective in the sense that it refers to objects, for by constructing facts it also constructs objects. It is also objective in the sense that it is intersubjective, for objects and facts depend on a consensus over the corresponding concepts and judgments. Science is not objective, however, in the sense that it describes, or makes assertions corresponding to, reality in itself, for objects and facts are constructions, not carbon copies, images, or icons of reality.

The relationship between objects or facts and reality deserves to be clarified a little further.

If we abandon real reference to its fate and adopt instead disquotationalism about reality, we are taken away from "realism about entities" just as disquotationalism about truth takes us away from "realism about theories."

Realism about entities claims that "a good many theoretical entities really do exist."[21] This view is ambiguous, though there is little doubt that by this realists mean to say not only that the terms of mature theories (putatively) *do* refer, but that we know to *what* they (really) refer. Realists would like to claim that our entities together with the properties we attribute to them are not constructs but genuinely existent bits of reality, or that successive entities represent ever deeper levels of reality. Our epistemology leads us to conclude, however, that this view is untenable.

We can of course agree that there may be many good reasons for one putative reference to be better than or preferable to another. This does not, however, authorize us to say that the best putative reference (including the one that has held out up to now against all criticism) is an existing entity in itself. This would be like posing an arbitrary limit to the advancement of knowledge, whereas we may discover one day that certain entities, previously considered real, no longer exist or no longer have the properties once attributed to them. Similarly, we are not authorized to conclude that the best putative reference corresponds to a deeper level of reality. This would, in some situations, be equivalent to ascribing incompatible entities and properties to reality. For example, the reality that corresponds to what we call an "electron" cannot be any deeper than the reality that corresponds to what Franklin called "electrical fire," because these two facts are incompatible, just as an entity with the properties of a fluid governed by fluid dynamics is incompatible with an entity with corpuscular and undulatory properties governed by quantum mechanics. Again, the reality that corresponds to what we call "light" cannot be, at ever deeper levels, first pressure or modification of a substance (Descartes), then movement of particles (Newton), then vibrations and oscillations of the ether (Maxwell), then an energy quantum (Einstein), then an entity with corpuscular and undulatory properties (quantum mechanics). Such vastly different ontologies simply cannot be used as building blocks for a single construction. A new conceptualization may well harmonize with a previous one, leaving its ontology intact, but radical conceptual innovations can, and often do, change everything. Nothing can guarantee that the ontologies of different theories will fit together like Chinese boxes or Russian dolls. Actually, the history of theories shows that an "epistemological break" often brings about ontological revision.

The realist of entities has one way out of these difficulties with a perfectly worthy argument: how can one doubt that certain theoretical entities of mature science are real, if they can be manipulated and if they can be used as instruments for carrying out certain operations? A case in point is that of electrons: "if you can spray them, then they are real."[22] According to this point of view, the existence of electrons does not depend on the success of the theories that are formulated in their terms, but on a "family of causal properties in terms of which gifted experimenters describe and deploy electrons in order to investigate something else, for example, weak currents and neutral bosons."[23] All difficulties of reference of theoretical terms would thus seem to be avoided. An electron is not the reference of a term within a theory, rather "electron" is the name of an entity. As McMullin writes, "the reason to affirm the entity's existence lies not in the success of the theory in which it plays an explanatory role, but in the operation of traceable causal lines."[24] To doubt the existence of an entity such as this would be like doubting that of a hammer whose causal lines are so easily traceable.

This solution makes such good sense that we are hard put to find fault with it. And yet, what does it prove? Strictly speaking, it proves that physicists carry out certain operations *o* on certain materials, with certain results *r;* that *r* is due to some entity contained in those materials; that in order to produce *r* thanks to the causal action of *o* this entity must have certain properties; that the name of this entity with such properties is "electron." Does this, then, prove that the electron is an element of reality in itself? It only proves—to use J. J. Thomson's words[25]—that "some unknown primordial substance *x*" which allows one to obtain certain results through certain operations *does* exist; but precisely *what* this substance *x* is does not depend on operations and results alone, it also depends on our theories. If these theories change, *x* can assume different values, even a zero value. The same is true for "hammer," though the word is admittedly less theoretical than "electron." Nobody doubts the reference of such a common household tool, with so many different uses. And yet, this reference depends on knowledge: in one form of knowledge, "hammer" may refer to one thing (and have its own name, say, "nail-hitter"), while in another, it may refer to another thing (and have a different name, say, "head-splitter"). And the two things may be totally, or partially, different entities.

Thus we are never completely sheltered from a potentially disastrous "meta-induction."[26] Rather than "substantial continuity in theoretical structures,"[27] the history of science shows substantial continuity in operations and results, that is, it shows that facts and objects that do not change through theory changes actually do exist. We shall return later to these facts—which we welcome as a gift of God—but we shall see that they do not always succeed in working miracles.

Let us return then to the question of the relationship between objects or facts and reality. As correspondence with reality has been lost, we are outside the realism of entities. But the negation of the realism of entities does not lead to antirealism any more than the negation of the realism of theories leads to idealism. If we know that, say, electrons exist, then we can say that "an electron is real if and only if it exists," following the disquotationalist scheme "*A* is real if and only if *A*." Similarly, if we know that the electron is real, then we can say that "'Electron' refers to electron," following the scheme "'*A*' refers to *A*."

How can this position be defined? As long as there is agreement on the main point, terminology is irrelevant. Now, the main point is as follows: common knowledge and science lead us to admit a world of entities (objects), such as cats, electrons, genes, etc., and events (facts), such as cats chasing mice, electrons rotating in their orbits, genes transmitting heredity, etc. If we admit these entities and events, then, on the one hand—considered in themselves, *ontologically*—they exist "out there," independently of our mind. But we cannot know of the existence of these entities and events except through the conditions we impose on them with our (I-, E-, C-) theories. Thus, on the other hand, if these entities exist, then—from our point of view, *epistemologically* or *transcendentally*—they are real according to the ways in which reality can be given to us, that is, they are real in the sense that they are the putative references of the terms we use to describe them. Outside these ways, entities are not real because we cannot grasp any meaning for real references.

Of all the labels in the philosophical market to describe this view,[28] the best remains Kant's "empirical realism," although Kant himself complained about being misunderstood.[29] Today, as then, it is not uncommon to find "hard" realists who, detecting a whiff of idealism in empirical realism, pull faces at it, turn their heads and make an effort to go on further. But, today as then, their efforts are doomed to failure.

The "metaphysical realist" ambitiously claims that "there exists one true and complete description of the way the world is," and that "truth involves some form of correspondence relation between words and thought-signs and external things and sets of things."[30] As we have seen, the obstacles in the way of this theory of truth are insurmountable.

The "critical realist" seeks refuge in a position that is humbler. He does not attempt to know *the real* and *the truth,* but to approximate to ever deeper *levels of reality* and to establish even greater *degrees of verisimilitude.* This is only seeming modesty, however, and nothing useful is achieved. The history of science shows that a level of reality achieved at time t can easily disappear at time t', just as the critical realist's own analysis shows that a degree of verisimilitude is no more than a presumption at time t that can change at time t'.

The "internal realist" tries to be more reasonable because he refutes things in themselves and levels of reality, and is happy with entities within some conceptual scheme. Nonetheless, he then goes on to add a substantialistic theory of truth that condemns his own attempts and reveals his hidden ambitions. For, on the one hand, if truth and reality are what one arrives at after an ideal rational justification, then, if such a justification cannot be achieved, truth and reality leave their place to *many* truths and *many* realities. This is not realism but more or less moderate idealism according to whether the conceptual scheme taken into consideration is more or less stable.[31] If, on the other hand, the ideal rational justification can be achieved (in the long run or at the limit of investigation), then reality and truth would no longer be within a conceptual scheme (in an ideal position all possible schemes would be exhausted) but become truth and reality *in themselves*. Thus, the internal realist starts out (well) as an empirical realist, but ends up (badly) as a nostalgic metaphysical realist.

The empirical realist tells all his ambitious colleagues that their unrestrainable desire to step out of themselves with substantialistic theories of truth can never be satisfied. His own realism, again, claims that entities and events of common knowledge and science exist independently of us but are known as putative references of terms and concepts and not as real references. Science can live "happily ever after" without such references.

6.4 Theory Change and Equivalent Theories

Objects and facts are highly in demand in the science market, but they represent a good which is hard to preserve, dependent as it is on the oscillations of consumer taste—even subject to extinction if there were simply no interest in buying. Leaving metaphor aside, what happens to objects and facts during a theory change? In order to compare two theories usefully, we need to know if they talk about the same entities.

Consider a predicate A = "electrical fire" whose reference is guaranteed in the way indicated in the previous section. We have:

$$< S, R, O, P, T_1 > = A$$

where T_1 is, say, Franklin's theory. Now let us replace T_1 with T_2, say the modern electron theory, and let B = "electricity." So we have:

$$< S, R, O, P, T_2 > = B$$

According to the disquotationalist view of reality, we say:

'A' refers to A = electrical fire
'B' refers to B = electricity

What we have to establish is whether A and B have the same reference. The answer can be found by examining the way in which results r can be obtained, according to T_1 and T_2, by carrying out operations o with instruments s, that is, considering the causal mechanisms m_1 and m_2 provided, respectively, by T_1 and T_2. If m_1 is compatible with m_2, in the sense that the ways in which the causal action exerted by entity A according to T_1 are the same as those of the causal action exerted by entity B according to T_2, or can be combined with them without any contradiction, then A and B have the same reference. If these ways differ, then A and B do not have the same reference and do not talk about the same entities. We can then lay down the following:

criterion of co-reference: two terms co-refer to the same entity if and only if the causal mechanisms with which they account for the same public, observable, or repeatable experimental situation, are the same or compatible with each other.[32]

In the case in question, it is unlikely that *A* and *B* co-refer to the same entity because the causal action of electrical fire (understood by Franklin to be a fluid) seems to conflict with the one that modern physics attributes to electricity (which has both corpuscular and undulatory properties and mechanisms). Since this question requires detailed description, we shall leave it aside. It suffices to note that a change of reference is possible, and to proceed to the further question of when this change takes place.

Let us take, as examples, the facts described by the following statements:

O_1: "An electron bombarded by particles leaps out of its orbit."
O_2: "Two bodies with contrary electrical charges repel each other."
O_3: "Venus has phases."
O_4: "Blood circulates."
O_5: "A group of Frenchmen attacked the Bastille in 1789."
O_6: "This rock falls."
O_7: "This needle moves from position *A* to position *B*."

In principle, all these facts could change. There are, however, different consequences to face and different prices to pay for this change. In some cases, change requires rejecting some E-theory or other: in the case of O_1, for example, one would have to rewrite a few chapters of atomic physics; in that of O_2, a few pages of electromagnetism. In other cases, a change would require a more radical rereading: to dismiss O_3, for example, a significant part of astronomy and optics would have to be thrown out of the window, while to reject O_4, a whole chunk of physiology, biology, and physics would go the same way. Finally, in still other cases, change would mean repudiating vast areas of our knowledge: to deny O_5, it would not be enough to alter a few textbooks, the whole history of the last two centuries would have to be rewritten. The amount of waste required by the change is a measure of the level of objectivity. The more wasteful the change, the more objective the fact. And a point arrives when the waste required is so great that its cost can no longer be afforded. In order to revoke the facts conveyed by O_6 and O_7, some I-theories must be modified, such as, for example, that the universe has directions, as well as some C-theories concerning the organization of our perceptions. Since C-theories are connected to our physiology and biology, these modifications are intolerable, inconceivable even, because they involve not only our knowledge but also our own apparatus for adaptation.[33] We can call facts such as these *facts with warranted credibility.*[34]

It is important to stress that not even these facts can be said to "correspond" to reality.[35] To claim that they do is not only senseless but also irrelevant. Let us take an example. Suppose that X = "This body falls" and Υ = "This body is resistent" are reports (facts) that guarantee our adaptation and survival; for example, in the sense that every individual who violated X and Υ by throwing himself off a balcony would die. We are thus guaranteed that X and Υ talk about reality and have a reference. But this is the putative reference that can change if certain E- and I-theories associated to X and Υ change (e.g., by changing the assumption that the universe has directions, X becomes false or senseless). The reference that does not change, the real reference, is unknown to us and this has no effect whatsoever on our adaptation and survival. Thus one cannot properly say that X and Υ "correspond" to a reality of things-in-themselves. The use of the notion of correspondence in this context would be nontechnical and misleading; "X corresponds to reality" here would mean "X is adapted to the environment" or "X is an ineliminable part of our way of being in the world." But the fact that something is adapted to the environment or is an ineliminable part of our way of being in the world, in the same way as a fish fin is suited to water, or teeth are suited to chewing food, does not mean that it corresponds to the environment in the sense that it depicts or reveals what this environment or world is like in itself. This would be possible only if the environment were determined, or materially generated, by our condition of adaptation, which is obviously to be excluded.[36]

So, the point is not whether facts with warranted credibility correspond to reality, because this cannot be ascertained, but that they are objective to the highest degree. Similarly the point is not whether facts with warranted credibility are absolutely elementary, pure, or simple, because they are not (even if they were it would be irrelevant), but that they are fixed, stable, and thus *shared*.[37] Our stratified view of theories and our epistemology allow us to combine the theory-ladenness of experience with its being stable and shared.

This result is precious but, unfortunately, not miraculous. First let me say why it is precious.

Suppose there is disagreement between upholders of two rival theories, T_1 and T_2. If they share a set of common facts, they can come to an agreement in order to set up a crucial experiment. Let us imagine a situation such as the following:

(I) $$((T_1 \land C \land H) \to O)) \land O$$
$$(T_2 \land C \land H) \to \neg O)) \land O$$

where C and H indicate, respectively, the initial conditions and the auxiliary hypotheses used to obtain the observational consequences (facts) O from T_1 and T_2. Clearly in this case the *modus tollens* does not allow us to conclude $\neg T_2$. Suppose, however, that during the controversy two rivals come to agree on both the initial conditions and auxiliary hypotheses. Can one in this case infer $\neg T_2$?

Quine maintains one cannot, for he claims that "any statement can be held true come what may, if we make drastic enough adjustments elsewhere in the system."[38] Just what adjustments must be made, though? Supporters of T_2 can recant certain theories T_3, T_4, etc. that prop up T_2, but in order to do so, several facts that support T_3, T_4, etc. also have to be revoked.

As we have already seen, such recantations are very costly, often prohibitively so. Moreover, in a dialectical exchange, adjustments made by one party are liable to be scorned by the other as ad hoc, absurd, artificial, or trivial, and thus to be disregarded or rejected. Although a move may be logically possible, it may still be dialectically unacceptable; therefore, although there are no logically or methodologically crucial eperiments, there may be *dialectically crucial* experiments. If, during a debate, a T_1 supporter promises to derive O while a T_2 supporter commits himself to $\neg O$, then if $\neg O$ is the case, T_1 is dialectically confuted in a crucial way. We shall be seeing an example soon.

But first let me say why our result is not miraculous. In order for a debate to be fruitful during a theory change, there must be contact not only between *supporters* of rival theories, but also between the *contents* of these theories. In other words, in order to express a preference between rival theories, facts which are *common* to and *relevant* for both theories must exist. During a theory change, however, it may happen that all the facts common to the two theories are irrelevant; and, vice versa, that all the facts that might be relevant are not in common.

Let me clarify this situation a little further. I shall call the conjunction of an E- and an I-theory the (heart of) a scientific system (or program). A theory change is a change in the whole system or in a part of it. More precisely, I call an *infra*-systemic change—that is, a change from an E-theory T_1 to another E-theory T_2 associated to the *same* I theory—an *ordinary theory change;* and an *inter*-systemic change—that is, a change from an E-theory T_1 to another E-theory T_2 associated to a *different* I-theory—a *radical theory change.*

I now say that a fact O *depends* on a certain theory if that theory attributes meaning (putative reference) to O according to the criterion of reference indicated in section 6.2 above. For example, "In this glass jar the electrical fire is condensed" depends on a theory like Franklin's, according to which electricity is a fluid of some kind. I also say that O is *common* to two theories if it has the same meaning (putative reference) in both according to the criterion of co-reference indicated in section 6.3. For example, "This glass jar attracts little bits of straw" does not depend on Franklin's nor on today's theory of electricity but is common to both. Finally, I say that O is *common* and *relevant* to two theories if it is independent of them (has the same meaning in both) and can be used to test both.

Let T_1 and T_2 be two E-theories and O a common and relevant predicate. Let us now take two I-theories, I_1 and I_2, associate them respectively to T_1 and T_2, and suppose a theory change takes place from $S_1 = (T_1 \wedge I_1)$ to $S_2 = (T_2 \wedge I_2)$. If I_1 and I_2 determine the meaning of O such that with the addition of I_1, O becomes O_1, and with the addition of I_2, O becomes O_2 and $O_1 \neq O_2$, then the fact O which was previously common and relevant for T_1 and T_2 will remain common but will no longer be relevant for S_1 and S_2, because the objects relevant for S_1 and S_2 are O_1 and O_2. If this holds true for every O for which S_1 and S_2 offer explanations, we no longer have the scheme (I) but the scheme:

(II) $((T_1 \wedge I_1) \rightarrow O)) \wedge O$
$((T_2 \wedge I_2) \rightarrow O)) \wedge O$, *for every common and relevant O.*

In cases such as these I say that S_1 and S_2 are *observationally equivalent.* One can see that two theoretical systems are equivalent when (a) they explain the same facts, (b) each translates these facts into its own language.[39] In our case, O is translated as O_1 in the language of S_1, and as O_2 in the language of S_2. I_1 and I_2 act as dictionaries.

Observational equivalence is an exceptional case that does not occur every time there is radical theory change. It only happens when, during radical theory change, an I-theory, associated with an E-theory, does not allow the use of facts that would be relevant to another E-theory associated with a different I-theory; or when, as Feyerabend says, "the conditions of meaningfulness for the descriptive terms of one language (theory, point of view) do not permit the use of the descriptive terms of another language."[40]

An aspect of the controversy between Luigi Galvani and Alessandro Volta and their followers concerning so-called "animal electricity" will help make things clearer.[41]

When the crural nerve and the corresponding muscle of a frog's leg are placed in contact with a mono- or bi-metallic arc, the leg contracts. Galvani explained this fact by introducing the theory of animal electricity (G) according to which:

G: There is a natural electrical imbalance between the frog's muscles and nerves.

Volta first opposed G with what I shall call the "*special* theory of contact electricity" (Vs) according to which:

Vs: Contact between two heterogeneous metallic conductors produces an imbalance in their electric fluid.

Then, when *Vs* was eventually confuted by Galvani, who obtained contractions *without* using metallic arcs by simply bending the crural nerve back onto the corresponding muscle, Volta introduced another theory, which I shall call the "*general* theory of contact electricity" (*Vg*) according to which:

Vg: Contact between any two heterogeneous conductors *whatsoever* produces an electro-motive force.

Let us now consider carefully the explanations that *G* and *Vg* offer for the frog's contractions. *G* says that the arc is passive and proves this by placing the frog's nerves and muscles into contact. *Vg* claims the arc is active and proves it by showing how two moist, heterogeneous conductors give signs of electricity. But what is the arc made of in Galvani's experiment, if not nerves and muscles? And if an arc made of nerves and muscles is active, as Volta claims, then nerves and muscles are vehicles of electricity, just as Galvani claims. Thus the two theories do not seem to be in opposition, for, as Galvani once suspected, "we're both saying the same thing, albeit with different words."[42]

And yet the two theories are in opposition and this has nothing to do with words. Since *G* considers nerves and muscles as specific *animal organs*, while *Vg* takes them to be *any moist conductors whatsoever*, *G* and *Vg* are opposed as regards their *concepts*. To be more precise, *G* and *Vg* are opposed because, associated to *G*, there is a *biologistic* assumption (or I-theory, say I_b), while associated to *Vg* there is a *physicalist* assumption (or I-theory, say I_p). Consequently, under *G* and I_b, the frog is an organic condenser and the frog's electricity is electricity *of* the animal (or *belonging to* the animal), while under *Vg* and I_p, the frog is a physical electrometer or battery, and the frog's electricity is electricity *in* the animal. The two theoretical systems are observationally equivalent.

Let us consider, for example, the following *O*-fact:

(1)

Under conditions $C_1, \ldots, C_n$, the frog contracts.

This fact is both common to and relevant for deciding between *G* and *Vs* because in the experiment without metals it confirms *G* and falsifies *Vs*, but it is useless for deciding between *G* and *Vg* because both theories can explain and predict it. In contrast, the *O*-fact:

(2)

> Nerve-muscle contact stimulates the frog,

which is relevant, and *might* decide between G and Vg, is ***not*** common to both because it depends on one of the two I-theories (assumptions) involved in the dispute, precisely on I_b. (2) is an O-fact only for G. The corresponding O-fact for Vg would be:

(3)

> Contact between the moist conductor nerve and the moist conductor muscle stimulates the frog.

This fact might also decide between G and Vg but it is not common to both because it depends on I_p.

As one can see, we find here a systematic procedure of bilateral translation. (2) is the translation of (1) according to I_b, while (3) is the translation of (1) according to I_p. (1) is the unvarying report while I_b and I_p are two dictionaries. The logical structure of the test of G and Vg is therefore the following:

$$((G \wedge I_b) \rightarrow O) \wedge O$$
$$((Vg \wedge I_p) \rightarrow O) \wedge O, \textit{for every common and relevant } O,$$

which is the same structure as scheme (II).

The above confirms some of our views about objects, reference, and the object-reality relationship. The predicate A = "animal electricity" can be defined by the quintuple:

$$< S, O, R, P, T > = A = \text{"animal electricity"}$$

where S is hands, O the bending of the crural nerve onto the muscle, R the contractions, P an element of $S \times O \times R$, and T a theory to be specified. When, in this quintuple, T is replaced by G and Vg, that is, by respectively Galvani's theory and Volta's general theory, and when the respective I-theories I_b and I_p are associated to each of these, we obtain:

$$< S, O, R, P, G, I_b > = A_1 = \text{"electricity } \textit{of} \text{ the animal,"}$$
$$< S, O, R, P, Vg, I_p > = A_2 = \text{"electricity } \textit{in} \text{ the animal."}$$

This illustrates various points. First, the meaning (putative reference) of a term requires theories ("animal electricity" is just a name with an ambiguous meaning unless the theory under which it is defined is not specified). Second, when theories change, references and objects can also change ("electricity *of* the animal" and "electricity *in* the animal" are two different objects). Third, between objects and reality-in-itself there is a gap that cannot be filled. A_1 and A_2 have putative references; their real reference is inscrutable.[43]

6.5 Rhetorical Strategies in Theory Change

In the previous sections I have tried to show that among the elements of the basis of scientific dialectics there are objective and common facts that in most cases allow a dispute between supporters of rival theories to be settled during a change of theory. In the worst possible circumstances, when the situation is desperate, the interlocutors can build themselves a minimal dialectical framework based, if necessary, on facts with warranted credibility. In other cases, though, the dispute has to turn to other factors of scientific dialectics, and in still others not even objective, common facts can be used since different theoretical systems can make up different objects and facts. In each of these cases different strategies must be followed, and a good scientist knows the best strategy to use to achieve the best results at the best moment.

A rhetorical strategy is an organization of argumentative techniques used to persuade an interlocutor. There are no fixed strategies because in science, as in war and play, strategy depends on the creativity and ability of the one who invents and practices it. There are however recurrent strategies, some of which have several variations. The following are the most frequent.

Strategy of crucial test. This strategy is used whenever there is agreement that the match between two theories can be regulated by shared facts considered to be conclusive. If T_1 foresees O and T_2 foresees $\neg O$, then if O happens to be the case and the conditions as well as the assumptions under which O is obtained are taken for granted, T_1 is proved. Since agreement with facts is the constitutive value of science, this strategy is the most efficient.

Strategies of empirical balance. Strategies of this kind also rely on facts but in situations in which they are not crucial. What count here are arguments showing that a certain theory satisfies certain requisites or achieves certain values, or that satisfies or achieves them better than a rival theory. There are many versions of these strategies, all familiar from standard methodological literature. For example, (a) T_1 is preferable to T_2 if it explains more facts; (b) T_1 is preferable to T_2 if it solves more problems; (c) T_1 is preferable to T_2 if it anticipates novel facts; (d) T_1 is preferable to T_2 if it explains facts that were anomalous for T_2; and so on.

When a scientist uses strategies like these, he will typically start by stating what the merit of his own theory is and then he will make every effort to show that it is one that he himself, or the whole community, or the most credited tradition approves of. He will go on to show that if one makes an objective assessment of the merits of rival theories at hand, his own will turn out to be the best.

Strategies of theoretical balance. These strategies are resorted to when facts are not considered sufficient, either because they are too few or because they are not thought to be the only testing ground. Usually they are employed when the empirical balance is equal or controversial. Their aim is to cast light on merits of other kinds, that is, merits regarding not just the relationship between theory and facts but also the internal quality of the theory and its external relations. These strategies also come in many versions. For example: (a) T_1 is preferable to T_2 if it uses fewer auxiliary hypotheses; (b) T_1 is preferable to T_2 if T_1 is consistent with other accepted theories and T_2 is not; (c) T_1 is preferable to T_2 if it is consistent with certain basic assumptions; and so on.

The task here is the same as in the above strategy. The scientist will have to select which internal and external advantages carry most weight for his interlocutors, and he will then highlight the fact that his theory is the one that possesses these merits to the greater extent. An able strategist knows how to touch the right chords in his interlocutors, and an able interlocutor knows how to avoid the dangerous moves and draw out the match by counter-opposing arguments that try to show that other merits must also be taken into consideration.

Dragging strategies. The name designates their effect: all strategies aim to convince, but while some (all the ones mentioned above) try to achieve this goal by heading straight for the thesis being questioned, others go by less direct routes. They do not take the main thesis to task but concentrate on another associated to it, so that if, during the debate the former is gained (lost), the latter is awarded victory (defeat).

These strategies are used not only when facts are not decisive or when they are irrelevant—as is the case with equivalent theories—but in all cases of radical theory change when sides must be taken in favor or against an assumption or I-theory. As usual, there are various kinds of "dragging" strategies.

The first kind is the "strategy of erosion." Preference must be expressed between T_1 and T_2 when they are associated to two different I-theories. This is impossible without deciding which I-theory is preferable, and this, in its turn, cannot be decided by empirical test. The attempt is thus to find some other merit of T_1 or T_2 so that one can be considered preferable to the other. If this is successful, the preferred theory will drag along the assumption associated with it. The rival assumption, strictly speaking, will not be falsified; it will be simply eroded, outgrown. As we have seen in the previous chapter, this was Galileo's strategy in tackling the assumption of the incorruptibility of the heavens.

Another dragging strategy is the "strategy of discredit." This follows the opposite path to that of erosion, in that it heads straight for the opponent's assumption and tries to demolish it. If this takes place, the theory linked to the assumption will appear less credible. Part of the Galvani-Volta controversy can be seen as an example of this strategy. Take, for example, Pietro Configliachi's preface to Volta's last memoir:

> Who today could be unaware that Physiology and Pathology, now more than ever, are in jeopardy from similar false suppositions; that they are using these as a base to create new erratic systems pertaining to organization, animal functions, and life, with no tangible benefit? On the contrary, such systems work to the detriment of young people, who, through no fault of their own, are still too inexperienced to guard themselves against the lure of novelty and all that is flashy and specious.[44]

Of course, discredit of this kind can succeed only if the community is willing to accept the values and spirit from which it emanates. In our case, if in 1814 a humble experimental physics professor was able to turn on physiologists with such arrogance, it was because he was confident that physics was recognized by the scientific community as being the dominant science.

Strategy of achieved or lost results. A strategy of this kind is at work whenever one praises a theory or assumption by suggesting that certain admired results would never have been achieved if that theory or assumption had not been chosen, or that certain promised results would be lost if that theory were not accepted. To say, for example, that a new line of inquiry opens new, previously unforeseen possibilities; or that what is now considered acquired knowledge would never have existed if a certain path had not been taken; or that certain unrelated facts could never have been arranged or would never have been explained if a certain theory had not been advanced or a certain assumption not been made, has the effect of casting favorable light on the theory upon which such admitted or supposed merits are recognized or insinuated to depend.

A typical example of this strategy is provided by Darwin. In the space of a few pages in chapter 14 of the *Origin of Species* ("Recapitulation and Conclusion") devoted to "special facts and arguments in favor of the theory,"[45] he uses the expressions "we see," "we can see," "we can understand," "we can easily understand," "we need not marvel," and other equivalents thirty-seven times; and phrases such as "facts are strange," "inexplicable," "utterly inexplicable" at least nine times, in order to show, respectively, that if one accepts the theory of natural selection, then a large class of facts ceases to be inexplicable, whereas if the theory that species are fixed is chosen, many things remain mysterious. The literary effect itself comes to support the strategy: what reader would be unmoved by such abundance? If at least some readers are converted—Hooker, for example, who became a "convert against his will," or Lyell, who became an "unwilling convert,"[46] just "half-hearted and whole-headed,"[47] or Huxley, who became a "bulldog"—then it is not hard to imagine that these readers will soon convince the rest of the community.[48]

All these strategies—some more than others—demand a certain mastery of rhetoric. Rhetoric is required for the strategy of crucial test because, as we have seen, in order for a theory to be crucially confuted its supporters must be induced to commit themselves during the debate to a particular result. Rhetoric is also needed for strategies of empirical and theoretical balance, which in the last resort depend on the choice of certain epistemic values and which are only efficient if these values—as well as their interpretation and their hierarchy—are shared and made explicit during the debate. Especially in a controversy between two important theories, this is rarely the case. Remember the cosmological controversy in which there are not enough available facts and other factors must be taken into consideration. The debate here rests on *which* factors to consider and *how* to interpret them. Bondi, for example, highlights simplicity or fruitfulness, but his critics retort that these values are less important than intertheory consistency. Or again, remember the controversy over the theory of natural selection. Darwin stresses the range of domain (the theory explains many facts of different classes), but his adversaries object that although this may be desirable it is by no means the most important point. As we have many times underscored, a balance of merits and demerits cannot be made with any instrument of precision, but only through the outcome of a debate. Finally, rhetoric is needed by dragging strategies because here everything depends on the choice of an assumption (I-theory) which, by definition, cannot be tested but only argued. It is also needed by the strategy of results, because here the acceptance of a theory or assumption depends on how appealing the achieved results, and how important the lost results, are depicted as being.

It might be objected that in the case of equivalent theories it is impossible, if not meaningless, to say that one is better than another. Therefore no rhetorical mastery would be needed because, as Quine has suggested, in cases such as these, "where there is no basis for choosing, we may simply rest with both systems and discourse freely in both, using distinctive signs to indicate which game we are playing."[49] In principle this is a wise recommendation but it is difficult to apply in practice, because, for example, to work on two equivalent theories at the same time may require too many intellectual and economic resources, or because the contenders may not discover immediately that the two theories are equivalent, or again because one may think that by extending the observational domain it would one day be possible to find some fact with regard to which the two theories behave differently. Quine's recommendation will at times be ignored: in the eyes of the upholder of one theory, the other theory, though equivalent, will not usually look very appetizing because of its ontology. For example, in the case of Galvani and Volta's theories, the biologistic ontology associated with the former arouses suspicion among the physicist upholders of the materialistic ontology linked to the latter.

To sum up, fixed rhetorical strategies do not exist; strategies are molded to the circumstances and invented (and a good inventor is already halfway to achieving his aim). But any strategy raises the same problem: if theory choice and acceptance depends on mastering rhetoric, in what sense can one say that the preferred theory is *better* than the one that has been laid aside? This is the problem of scientific progress that I shall now deal with.

Seven

Rhetoric and Scientific Progress

Scientific advances are so difficult to achieve that every useful stratagem must be used.

W. I. B. Beveridge, *The Art of Scientific Investigation*

7.1 The Problems of Scientific Progress

Galileo wrote that science "can only improve."[1] Others, such as Sarton[2] and Popper,[3] have gone further and claimed that science is the *only* intellectual endeavor in which it makes sense to talk of progress. Although the latter view is open to debate,[4] we can take the former as a basic intuition: in every theory change, a subsequent theory T_2 is (and is considered to be) progressive in relation to a previous theory T_1. To flesh out this intuition, two problems must be dealt with: what does "T_2 progresses over T_1" *mean*? and how can we *determine* that T_2 progresses over T_1? These two problems parallel the two problems about truth: the first concerns the definition of progress, the second our ascriptions (or determinations) of progress. I shall try to show that the way we dealt with the problems of truth also applies (*mutatis mutandis*) to the problems of progress.

Let us begin with the definition. Generally speaking, the idea of progress contains two others: change and improvement. Change raises no special difficulties, but what precisely is meant by a subsequent theory making an improvement on its predecessors, needs to be clarified. To start with, a theory can be said to make an improvement on another if it is comparable with it, which requires two conditions to be satisfied.

The first condition is that, during the change from T_1 to T_2, there must be some continuity as far as the objects (facts, problems) that T_1 and T_2 deal with are concerned. It makes no sense to say that Dalton's theory of atoms is progressive over, say, Leibniz's theory of monads, while it makes perfect sense to say that J. J. Thomson's theory is progressive over Dalton's. In short, one can say T_2 makes progress with respect to T_1 only if there is *continuity of domain* between T_1 and T_2.

The second condition is that there must be some continuity as regards the values by which T_1 and T_2 are oriented. Suppose T_2 aims at explaining certain phenomena of its domain, while T_1 intends using them for emotional, aesthetic, or didactic effect. In this case one clearly cannot say that T_2 is progressive over T_1. To give an example, it would be as absurd to say that Weinberg's *First Three Minutes* makes progress over Genesis as it would be to claim that a cosmological textbook is progressive over a science-fiction novel. In short, one can say T_2 is progressive over T_1 only if, in addition to continuity of domain, there is also *continuity of values.*

Obviously neither of these conditions would be satisfied if the relationships between the values V and domains D of two theories were of the kind $V_1 \cap V_2 = \emptyset$ and $D_1 \cap D_2 = \emptyset$. Fortunately, this cannot be the case in science. If two theories are *scientific,* then they must at least share the constitutive value of science (agreement of cognitive claims with facts); and if they are *competing,* then they must be homogeneous,[5] that is, they must share a minimal core of facts. Even in the case of those radical theory changes in which the references of the theoretical terms of two theories (such as those of Galvani and Volta examined in the previous chapter) becomes incommensurable, the two theories still share certain elementary facts and certainly have the facts with warranted credibility in common. That two such theories construct different theoretical entities out of the same observational and experimental material, does not mean they have no common ground. Indeed, without it, theory change and consensus shifts from one theory to another could never have taken place.

To say that two scientific theories T_1 and T_2 exhibit some continuity of domains and values it is enough to say they belong to the same tradition and can therefore be compared. But once such continuity has been specified, in order to be able to state that T_2 makes progress over T_1, one must establish precisely in respect of *which* domain and value T_2 makes progress over T_1. For example, Copernicus' theory is progressive over Ptolemy's in respect of simplicity; Volta's theory can be said to be progressive over Galvani's in respect of the domain of physical entities, and so on.

Let us consider the question from the point of view of values. Two solutions compete here. The first maintains that scientific progress is relative to the values holding in, or admitted by, the culture of the time at which the theory change takes place. The second claims it is relative to the values of our current culture.[6] According to the former view, scientific progress can only be internal (to a culture, paradigm, research tradition, etc.), while according to the latter, it is continuous. The first solution stems from the (democratic) view that each culture is, and is to be considered, a world of its own, and that it makes no sense to compare one culture with another, let alone to place them on a hierarchical scale. The second solution grows out of the (Whig) view that cultures can be compared and located along the same line of advancement. For the first view, what follows is merely a different episode from what precedes; for the second, it is anyway an improvement, a step forward.

I have implicitly taken a position on this matter when dealing with the question of relativism. It may be (though it is far from obvious, if not highly questionable) that in politics, ethics, religion, or related areas the democratic view is more attractive; after all, to be democratic and tolerant towards a different culture is fashionable and costs nothing as long as one is not required to take up arms in a confrontation between that culture and one's own. And yet in the history of science one can only be a Whig. When reconstructing past theories, the historian of science, in order to consider them *scientific,* cannot but take them as being oriented towards epistemic values. But how can he know that certain values pursued by past researchers are epistemic? He cannot simply rely on their professed commitments, for if these prove they have pursued, say, *A, B, C,* etc., the historian of science must have some preliminary idea as to what counts as an epistemic value in order to establish whether *A, B, C* are epistemic or not. And since this idea can stem only from his present-day values, the historian of science is obliged to use his own values as a point of reference, thereby placing scientific theories within a single tradition whose roots lie in the past and reach to the present day.

7.2 Progress, Improvement, and Victory

Let us then take the Whig view, and decide to appraise progress in respect of our own values. One can ask: what values?

There are conflicting opinions on this matter, as shown by the list below, which contains the main models of progress discussed in the literature.

1. $T(T_1) \subset T(T_2)$
2. $Q(T_1) \subset G(T_2)$
3. $Tr(T_1) < Tr(T_2)$
4. $CN(T_1) < CN(T_2)$

5. $e(T_1) < e(T_2)$
6. $p(T_1) < p(T_2)$

Each model in this list is based on a definition of progress according to the following schema:

Scientific progress means an increase in *x* (or *y*, or *z*, etc.)

where *x, y, z,* etc. are properties of theories considered as values: for example, truth (*T*);[7] answered questions (*Q*);[8] truthlikeness (*Tr*);[9] confirmed novel facts (*CN*);[10] problem-solving effectiveness (*e*);[11] and the quantity and importance of solved problems (*p*).[12]

All these definitions present well-known analytical difficulties concerning the explication of the terms with respect to which progress is appraised. It is not entirely clear, for example, what the truths discovered by a theory are, whether and how the truthlikeness of a theory can be appraised, what amounts to a correct solution of a problem, and so on. Above all, it is not clear why progress should be defined in terms of one value rather than another.

It would be an easy exercise to show that, for any given model, there is an episode of theory change that exemplifies it. However, to interpret the history of science, or to consider scientific practice, as always honoring the same single value or set of values would be like putting them into a straitjacket, since not only do scientists replace a preceding theory with a subsequent one at certain times on the grounds of certain values and at other times on the grounds of others, but they also reserve to themselves the right to do so. The reasons why Galileo's astronomy replaced Ptolemy's are not the same as the reasons why classical mechanics was replaced by relativistic mechanics and fixism by evolutionism; nor would astronomers or biologists commit themselves to a fixed set of reasons.

In contrast, to claim that there is an all-embracing value which, like a kind of highest common denominator, applies to all occurrences of theory change would be like draping a baggy jacket over the history of science, since such a value could only be highly generic. On this point, we are faced with the same difficulties of historical methodologies we met in chapter 2. For example, to declare that subsequent theories are always more "reliable" than their predecessors[13] is very vague, since there are many indicators of reliability and they can and do change. Equally, nothing very illuminating is revealed by claiming that the history of science shows that new theories "solve some of the problems unsolved by their rivals."[14] The same holds if progress is defined in terms of the constitutive value: true, all scientific theories must "agree with facts," but what kind of agreement? What facts? Although the constitutive value is invariant and represents the minimum core around which every theory change takes place, it must be combined with other epistemic values and is anyway open to different interpretations.[15] If these combinations and interpretations are rendered very precise, it may happen that the resulting model no longer covers episodes we would like to consider progressive. This is precisely the same situation that led us in chapter 1 to the paradox of scientific method.

Rather than pursuing any further the attempt to find a single model of progress, it seems more reasonable to adopt a pluralistic attitude and admit that no model of progress can serve all purposes and save all episodes of theory change. The fact remains, however, that every theory change constitutes an improvement when judged with respect to the values of subsequent theories and of our present theories which are their successors. As an answer to our first question, then, I shall offer the following minimal, deflationist definition:

(P) T_2 is progressive over T_1 if and only if T_2 is better than T_1.

This definition forswears analyzing the idea of progress in terms of specific relationships between certain fixed properties of theories. It is like the minimal definition of truth which also forswears analyzing the idea of truth in terms of certain substantive properties of propositions. Although the definition may seem trivial, it does not give up the idea of progress, nor does it lead us to renounce our intuitions concerning the link between science and progress. Just as with the minimal definition of truth, the idea that science aims at truth is not lost because it is replaced by the idea that it aims to know such things as whether snow is white, planets orbit elliptically, etc., so with the minimal definition of progress, the question whether science progresses is not given up because it is replaced by such questions as whether Galileo's theory is better than Aristotle's, Newton's better than Galileo's, etc. In the same way, just as the minimal definition of truth leaves us with the problem of assertibility conditions, the minimal definition of progress leaves us with the problem of improvement conditions.

This is the second question about scientific progress that now crops up: if progress is made in respect of different values at different times, how can we determine each time that one theory is better than another?

The standard answer is methodological: rules exist that determine, on the basis of certain properties, whether a theory is progressive over its rivals. This is the Cartesian view, or its neo-Cartesian version, depending on whether such rules are considered permanent in every theory change or are simply agreed upon at a given juncture.

The second answer is cynical. It denies there are methodological rules for determining objectively whether one theory is better than another and limits itself to recording that at a certain, given moment a certain, given theory has proved to be successful for some reason or other and has marked a victory over its rivals. Its upholders then claim they have made an improvement simply because they define progress in terms of following the victorious theory. According to this view, all there is to say about progress (as well as about rationality and truth) is to describe the theory change and take note that the new theory dictates the new rules of appraisal. For example, one can say that Galileo is better than Bellarmine because he "just lucked out," not because his theory came any closer to, or had any better insight into, nature.

Here too the two views prove to be the two sides of the same Cartesian coin, or the two horns of the same Cartesian dilemma: either our determinations of progress are objective (for example, in the sense that we can prove that one theory reveals deeper levels of reality than its rival does), or they are deceptive (for example, because there are no such things as ultimate reality, absolute truth, and so on, to which theories are supposed to approximate). By rejecting this coin, and adopting the dialectical view, the problem of the determination of progress takes another path: it too recognizes that methodological rules for expressing preference do not exist, and links progress to success, but denies there is nothing more to say about the reasons for this success. The crucial notion to concentrate on is that of victory.

To maintain that the best theory is the one that has won is clearly unsatisfactory if we do not specify *why* it has won. One the one hand, we obviously cannot claim that anyone who happens to win a debate, no matter how, for this reason alone gains the right to declare that his own theory represents progress over his rival's. This would be too cheap. On the other hand, we cannot even say progress has been made when the victory is definitive and leaves no room for discussion: this would be tantamount to setting a limit to the fallibility of science and clipping the wings of its growth. What we would like to say is that only he who has achieved an *honest victory,* at least a *non-Pyrrhic victory,* has the right to claim progress has been made. But what is an honest victory?

This question seems to lead us back to a dilemma. If we define as honest a victory secured through abiding by a certain set of permanent or agreed upon rules, we fall into the methodological view. If we deny the existence or efficacy of any set of this kind, and fix no constraints, then an honest victory becomes just a victory, and we fall into the cynical view. Yet again, the dialectical view intends avoiding the two horns of this dilemma: according to this view, the victory of a theory is honest if the supporters of that theory produce stronger arguments than those advanced by supporters of its rivals.

There is a marked difference between the dialectical view and the others: while the methodological view links victory to the observance of certain requisites fixed by a code acting as impartial arbiter, and the cynical view rejects both the arbiter and the objective justifications of victory, the dialectical view links victory to the argumentative capacity of competitors. For these reasons, while the methodological view is committed to defining the concept of *impartial arbiter,* an arbiter independent of the specific trial he is called in to judge, the dialectical view is committed to defining the concept of *honest victory without an impartial arbiter,* a victory to be adjudicated by no other resources than the ones offered by the discussion. Before exploring this second view, I shall try to show that, even in the matter of progress, the first view cannot be maintained.

7.3 Victory and Rules of Preference

The methodologist's problem is to hang the determinations of progress onto the pegs of certain requisites fixed by certain rules. Kuhn's idea is that this is bound to fail. Laudan—whom I shall consider, as I did in chapter 2, an important exponent of the methodologist view—has claimed it will not. In order to take sides, we must see more clearly how they disagree.

Kuhn does not deny there are a "set of commitments without which no man is a scientist,"[16] or good reasons for making a choice,[17] or a desideratum such as accuracy of fit to nature that is constitutive of science.[18] In his turn, Laudan does not deny that these desiderata may change over time, or may be interpreted in different ways. What he does deny is that subjective or pragmatic factors that cannot be made superfluous or uninfluential by an appropriate methodological rule enter into theory preferences and determinations of progress. The crucial point of disagreement concerns that "mixture of objective and subjective factors, or of shared and individual criteria" Kuhn maintains cannot be, and Laudan claims can be, dissected,[19] that is, those shared criteria of appraisal that Kuhn considers "not by themselves sufficient to determine the decisions of individual scientists" and therefore to be "fleshed out in ways that differ from one individual to another,"[20] and that Laudan considers "sufficiently determinate that one can show that many theories clearly fail to satisfy them."[21]

The idea of an impartial arbiter serves precisely this purpose: he is called to desubjectify theory preferences. For example, if disagreement arises over which theory solves the most important problems, the methodologist requires a distinction to be made between subjective importance and "epistemic or probative importance" in order to "desubjectify the assignment of evidential differences by indicating the kinds of reasons that can legitimately be given for attaching a particular degree of epistemic importance to a confirming or refuting instance."[22] Note that behind this requirement there lurks the Cartesian syndrome. The methodologist implicitly argues that if decisions were not dictated by rules, they would depend on personal ingredients and would not therefore be rational. As Laudan writes, "the *rational assignment* of any particular degree of probative significance to a problem must rest on one's being able to show that there are viable *methodological and epistemological grounds* for assigning that degree of importance rather than another."[23]

In order for a rule to be able to determine a theory preference or a verdict of progress, the holistic picture of theory change must first be dissected. This is necessary because, as we have seen, if there were no overlapping between the domains, standards, and values of theories we could not even talk in terms of *scientific* change. The reticulational model with which Laudan substitutes this picture, however, does not seem to live up to its aims. Take a collective C_1, comprising certain methods M_1, certain aims A_1, and a theory T_1, with the following relationships: "A_1 will justify M_1, and harmonize with T_1; M_1 will justify T_1 and exhibit the realizability of A_1; and T_1 will constrain M_1 and exemplify A_1."[24] Now introduce a theory T_2 into C_1, the first step for moving to C_2. Laudan thinks that "the rules M_1 will be consulted and they may well indicate grounds for preferring T_2 to T_1."[25] But this seems hardly likely, because the reticulational model itself shows the *T-M* pair is firmly linked,[26] and, anyway, in the most important cases M_1 does not permit a preference for T_2. To return to the cases of theory change examined in chapter 3, Aristotle's method excludes Galilean physics, Bacon's method is at odds with the theory of natural selection, and inductive method is an obstacle for the cosmology of the steady-state. When a theory change is very deep, the new theory is associated *from the outset* with a new method.

The methodologist might well object that since there is a certain continuity of values and domains between different theories, their supporters must also share some rules. Let us grant this is the case and take Laudan's favorite rule:

> *R*: "Prefer that theory which comes closest to solving the largest number of important empirical problems while generating the smallest number of significant anomalies and conceptual problems."[27]

Suppose now there is conflict between the supporters of two rival theories concerning which is preferable. Resorting to *R* will only settle the controversy if the contenders interpret the key terms of *R* ("important problem," "significant anomaly") in the same way and if they apply *R* in the same manner. But this is precisely the heart of the conflict. Since nobody has more authority than any other to provide authentic interpretations and correct applications, only a debate can determine the preference decisions.

Let us see how. If an interlocutor A supporting T_1 aims at confuting another interlocutor B supporting T_2, he must start finding a common ground of agreement. Suppose both come to accept problem-solving effectiveness as the main test for theory preference. If during the debate B concedes that a certain problem p is important, that solving p means deducing a certain fact f from a theory T_1, and that f follows from T_1 but not from T_2, then A confutes B thanks to an argument based on B's own concessions. It is not, therefore, rule R that decides whether T_2 is preferable to T_1, but the debate that establishes that A's arguments are stronger than B's. Within this debate, R functions as a commonplace of preference ("those theories that solve important problems are to be preferred"); it is an *éndoxon*, a reputable opinion shared by both A and B and admitted by the community.

The methodologist might still object that in certain cases methodological rules exist that determine theory preferences univocally without needing to resort to a debate. For example, a rule such as "Prefer those theories that are internally consistent" is so clear that given any two rival theories, it is easy to decide which is preferable, with no ambiguity. Laudan is right to insist that sufficiently precise rules exist, and if Kuhn claimed that all rules underdetermine theory preference he would be making a mistake. But precise rules determine decisions when they become an explicit commitment made by the contenders in a debate. A debate is then always needed because it is through a debate that the contenders' commitments come to light. Outside a debate there is no rule; inside the debate, a rule works like a value and is one of the factors on the basis of which the discussion and confutation takes place. Methodological rules of preference for determining progress thus vanish and are replaced by the outcome of the debate conducted according to the factors of scientific dialectics. From the model of the game with three players we know that these factors guarantee a victory without an impartial arbiter. How this model solves the problem of our determination of progress and what picture of progress stems from it, still needs to be examined.

7.4 Honest Victory without an Impartial Arbiter

According to the view maintained here, for every theory change there are reasons as to why one theory is replaced by another, even though these reasons are not always the same. The image of scientific progress stemming from this view, then, is like the one advocated by Kuhn: if at every theory change improvement can take on a different guise, the series of scientific

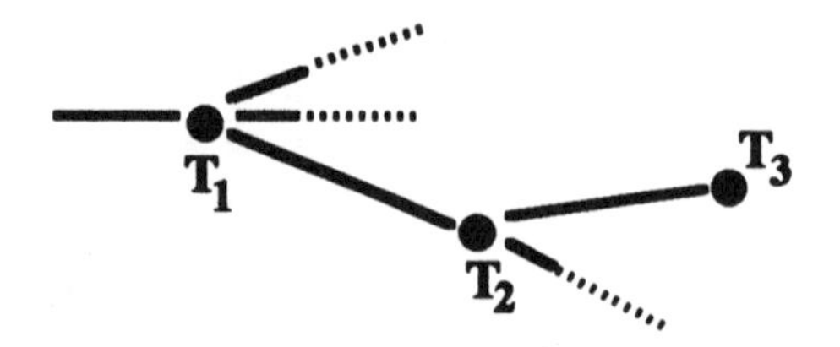

Fig. 7.1.
The tree of scientific progress.

theories is an evolutionary tree marked at intervals by knots which are followed by branches (see fig. 7.1). The line has a point of departure which is the situation of knowledge at a given time, but it has no established itinerary and no predetermined destination. The only thing that can be said is that any later step along the line constitutes an improvement over an earlier one.

Each knot represents a difficulty encountered by a theory within the area of the configuration of substantive factors of scientific dialectics holding at the time. For a theory, this configuration is something like an ecological niche for a species. If it changes, a selective pressure is exerted on the theory. For example, a new fact that is incompatible with the existing theory may appear, an inconsistency between the existing theory and another well-founded theory may emerge, a new consequence of the theory may clash with already accepted knowledge, a fundamental assumption may be doubted, a supporting value may change position in the hierarchy accepted at the time, and so on.

Each branch of the evolutionary line is an answer to the pressure on the existing theory and represents an attempt to solve the difficulty, comparable to a variation or incipient species: it could be an adjustment of the old theory, for example, a modification obtained with the introduction of an auxiliary hypothesis; or a new theory obtained, for example, with a change in some basic assumption. In any case, a debate arises among the supporters of different proposals. We know how it goes: the contenders must first find, among the substantive factors of scientific dialectics, a minimal area of agreement (the *éndoxa*); on the basis of this area and their mutual concessions, they must then produce arguments with the aim of confuting each other. The debate is (rationally) over when the arguments of one party are stronger than those of the rival party. At that point the winning party has the right to consider its own theory progressive over its rival's theory according to the following schema:

> If the arguments supporting T_2 are stronger than those supporting T_1, T_2 is progressive over T_1.

Note that "the arguments supporting T_2 are stronger than those supporting T_1" does *not* mean "T_2 is progressive over T_1". The schema above does not define progress (whose definition is provided by *P*), it only offers a presumption of progress. This may raise immediate objections. The first concerns the configuration of the factors of scientific dialectics which is responsible for the presumption of progress: on the basis of which configuration should arguments be considered "stronger"? The second regards the relationship between success in a debate and the progressive nature of a winning theory: is it not perfectly possible for a party to lose (win) a debate even when its own theory is better (worse) than its rival's?

I have already provided an answer to the first objection. Even though the configuration of factors of scientific dialectics may change during a theory change, the continuity that exists among these factors is enough to guarantee a debate. Those arguments that confute a rival on the basis of, or starting off with, a minimal area of agreement are stronger. Locating that area depends on the interlocutors' rhetorical abilities. Though it may not be easy if their opinions are far apart, it is not impossible, for even if a new opinion is radically innovative, there is no way it can alter or reject at the same time all the factors of scientific dialectics without being cast outside scientific tradition itself.

As regards the second objection, it can be admitted that one may lose (win) a debate for a theory even when there are good (bad) reasons for preferring that theory. Nevertheless, only the continuation of the debate, or a new debate, can establish whether this has taken place. Even the latest theory, which is taken as the best and in the light of which previous theories are judged, is the product of a debate. To try and place oneself outside a debate is like trying to find a vantage point from which one can perceive whether a theory is "really progressive" or "really true"; that point simply does not exist, and makes no sense.

A few comments are needed.

According to the view maintained here, progress is not an approximation to a fixed goal. Although a subsequent theory is closer to our values, these values are those of the theories that are successful at present, they are not the unique and permanent goals towards which the series of theories aims. As a matter of fact, the series of theories has no external goal to be reached; its only aim is internal, and that is improvement, which is always guaranteed because for one theory to replace another it has to win a debate with the other and must therefore for some reason or other represent an improvement on it. Galileo was then right when he claimed that science "can only improve."

Galileo was wrong, however, to think that improvement is always cumulative, like a geographical exploration in which previously conquered territories are preserved. Improvement can take on different forms at every theory change. That there is a certain continuity among these forms, does not mean that the winning theory in a debate contains all its predecessors' successes, because the configurations of factors change so that the reasons leading in a certain direction at one knot are not necessarily the same at another. It may then happen that a theory explaining some facts and failing to explain others is replaced by another theory that does not explain all the facts explained by its predecessor but is nonetheless better in another sense. The phenomenon of "Kuhn losses" in theory change does not mean improvement is not progressive, only that it is not cumulative.

Even if progress is not cumulative, however, it is knowledge-extensive. At every knot in a series of theories, different directions branch off, each of which opens up a range of possibilities. A subsequent theory must solve at least some of the problems of its predecessor. It can do so in various ways, for example, by explaining a recalcitrant fact, unifying separate domains, or changing some deeply rooted assumption that reveals itself to be the cause of the knot. In each case, in the light of the new theory one can see *why* there were difficulties that paralyzed the previous one and understand *how* such difficulties can be overcome. This amounts to extending knowledge, conquering new facts. Anti-representationists like Rorty are quite right when they say this does not mean the new theory is a more faithful mirror of the deep constitution of nature. Yet, if the new theory is better, an old picture is modified or canceled, a new one is given, and our knowledge is extended.

Note that, although it maintains that the best theory is the one that wins, the dialectical view is not cynical. "Progressive" is not predicated of the theory of those who happened to win a debate but of that theory of those who won the debate after a rational discussion based on the substantive factors of scientific dialectics. To stress the point once again, this victory is *honest,* because such factors are available to all the interlocutors. Moreover, it is *without an impartial arbiter* because it does not depend on the fact that the winning theory possesses certain characteristics predetermined by a methodological code, but on the fact that the debate shows it has advantages its rivals lack. And in the debate, nobody and nothing is impartial: neither the rival interlocutors who are parties to the case, nor nature, which is a witness subjected to the influence of those who interrogate it.

It remains to be said that the dialectical view unifies two apparently contrasting needs. As we have seen, Laudan wants to save the objectivity of theory preferences and maintains that, at least in some cases, rules exist for determining them univocally, while Kuhn wants to save the role of the individual and maintains that these rules are not in themselves sufficient.

If the determinations of progress are made dependent on a debate conducted on the basis of the factors of scientific dialectics, both needs are satisfied. The determinations of progress are desubjectified, not because they derive from rules working as algorithms or yardsticks, but because they depend on the strength of the arguments used in the debate. Furthermore, subjective factors come into these determinations, not because they directly influence decisions, but because the choice of arguments, case by case, is up to individual scientists. One can say the determinations of progress are desubjectified, not because they have been mechanized or even socialized (that is, imposed by external social or cultural factors), but because they have been *intersubjectified*.

We thus return to the typical connotation of the dialectical model of science. The game with three players allows us to do away with method without losing rationality, objectivity, truth, and progress. There is a cure for the Cartesian syndrome after all.

Notes

Introduction

1. Bacon (1620a), p. 109, #122; cf. also pp. 62–63, #61: "but the course I propose for the discovery of sciences is such as leaves but little to the acuteness and strength of wits, but places all wits and understanding nearly on a level."

2. Descartes (1628), p. 11.

3. Ibid., p. 17. On the reason-method relationship in Descartes, see Schouls (1980), esp. chap. 3.

4. Descartes (1628), p. 16.

5. Leibniz (1961), 4:329.

6. Ibid., 7:202.

7. Ibid. p. 237.

8. I introduced the idea of a Cartesian syndrome in my 1982 work. I came then to appreciate the obscure ways of serendipidity reading the following passage from Bernstein (1983), p. 18: "with a chilling clarity Descartes leads us with an apparent and ineluctable necessity to a grand and seductive Either/Or. *Either* there is some support for our being, a fixed foundation for our knowledge, *or* we cannot escape the forces of darkness that envelop us with madness, with intellectual and moral chaos."

9. Eysenck (1953), p. 226.

10. Kant (1783), p. 4.

11. Popper (1959), p. 39. Later, Popper weakened this point of view. See (1963), p. 257: "the criterion of demarcation cannot be absolutely sharp"; (1974), p. 981: "any demarcation in my sense *must* be rough," and p. 984: "such a rule of method is, necessarily, somewhat vague — as in the problem of demarcation altogether."

12. See, respectively, Lakatos (1974), p. 315n; (1976), p. 168 and p. 192.

13. Lakatos (1970), p. 31.

14. Ibid.

15. Lakatos (1971), p. 103.

16. Lakatos (1970), p. 91.

17. For a more detailed examination of Lakatos's Cartesian project and its failure, see Pera (1989a).

18. Feyerabend (1975), p. 23.

19. Ronchi (1983), p. 214.

20. Ibid., pp. 236–37.

21. Ronchi (1978), p. 866.

22. Ibid., p. 866.

23. Ibid., p. 867.

24. Feyerabend (1975), pp. 153–54.

25. Feyerabend (1970), 2:150.

26. Ibid., p. 11.

27. "None of the methods which Carnap, Hempel, Nagel, Popper or even Lakatos want to use for rationalizing scientific changes can be applied, and the one that can be applied, refutation, is greatly reduced in strength. What remains are aesthetic judgements, judgements of taste, metaphysical prejudices, religious desires, in short, what remains are our subjective wishes" (Feyerabend 1975, p. 284–85).

28. My expressionistic technique may not do full justice to Feyerabend, because, at times, he seems to attribute rationalists with the Cartesian dilemma. He writes, for example (1978, p. 14): "And as rules and standards are usually taken to constitute 'rationality', I inferred that famous episodes in science that are admired by scientists, philosophers and the common folk alike were not 'rational', they did not occur in a 'rational' manner, 'reason' was not the moving force behind them and they were not judged 'rationally'." However, the fact that Feyerabend has often taken "tricks," "irrational means," and "propaganda" as alternatives to method, still convinces me that he too is affected by the Cartesian syndrome. If I am wrong then my criticism is to be addressed not to Feyerabend in flesh and blood, but to Feyerabend as a character in search of an author.

29. Rorty (1982), p. 192.

30. See respectively, Rorty (1979), p. 318, and Rorty (1982), p. 195.

31. Rorty (1982), p. 193.

32. Ibid.

33. Rorty (1979), p. 331.

34. Ibid., p. 318.

35. Rorty (1982), p. 192.

36. Bloor (1976), p. 37.

37. Hesse (1980), p. 33.

38. Kuhn (1962), p. 150.

39. Ibid., p. 158.

40. Ibid., p. 151.

41. Ibid., p. 94.

42. Ibid., p. 152.

43. Ibid., p. 152.

44. Ibid., pp. 157–58.

45. Ibid., p. 94.

46. Ibid., p. 159.

47. Ibid., p. 147.

48. See, in particular, Kuhn (1970a), in which he wrote, "to name persuasion as the scientist's recourse is not to suggest that there are not many good reasons for choosing one theory rather than another" (p. 261); see also Kuhn (1970b), in which he stated that his position did not imply "either that there are no good reasons for being persuaded or that those reasons are not ultimately decisive for the group" (p. 199). See, finally, Kuhn (1977), chap. 13. The difficulty in reconciling Kuhn's new views with previous statements is discussed in Shapere (1971) and, with different results, Doppelt (1978) and (1983).

49. Here are some typical (Cartesian) reactions to Kuhn's book: "It follows that the decisions of a scientific group to adopt a new paradigm cannot be based on good reasons of any kind, factual or otherwise" (Shapere 1966, p. 83); "The general conclusion to which we appear to be driven is that the adoption of a scientific theory is an intuitive or mystical affair, a matter for psychological description primarily, rather than for logical and methodological codification" (Scheffler 1967, p.

18); "Kuhn's premises (that scientists do not always change paradigms, or fail to change paradigms, on rationally justifiable grounds) fail to support his conclusion (that there are no rationally justifiable grounds for paradigm change)" (Purtill 1967, p. 58); "In Kuhn's view, scientific revolution is irrational, a matter for mob psychology" (Lakatos 1970, p. 131): "Some historians and philosophers (e.g. Kuhn and Feyerabend) have argued not merely that certain decisions between theories in science *have been irrational,* but that choices between competing scientific theories, in the nature of the case, *must be irrational*" (Laudan 1977, p. 3). For a very different reading of Kuhn's book, see Doppelt (1978) and (1983).

One

1. *Phaedrus* 268a–c.
2. Ibid. 270d.
3. Ibid. 271b–c.
4. Ibid. 264b.
5. Ibid. 264e.
6. Ibid. 264c.
7. The correct succession of steps that Plato calls "dialectical procedure" (266c) is as follows: "the first is that in which we bring a dispersed plurality under a single form, seeing it all together"; the second is "the reverse of the other, whereby we are enabled to divide into forms, following the objective articulation" (265d–e). In other words, the two stages are collection and division.
8. The first meaning of "method" noted by Bonitz in Aristotle is that of "*via ac ratio inquirendi.*" See Bonitz (1955), under the entry *methodos.*
9. This is the meaning of scientific method adopted, for example, by Beveridge when he stated that "the phrase 'the scientific method' can only be taken as meaning the procedures and mental processes used by scientists in advancing knowledge." See Beveridge (1980), p. 54.
10. See extract quoted in Introduction, p. 2.
11. See extract quoted in Introduction, p. 4.
12. See Popper (1959), p. 49, where methodology is defined as the theory of the "rules of scientific method."
13. See Lakatos (1971), p. 103, where he states that "modern methodologies or 'logics of discovery' consist merely of a set of (possibly not even tightly knit, let alone mechanical) rules for the appraisal of ready, articulated theories."
14. The difference between the second and the third meaning can be clarified with a couple of examples. Consider the case of clinical medicine. Method, in the second sense, could include the rule that no diagnostical hypothesis is acceptable if it is not independently backed up by facts, while in the third sense it specifies the way in which the operations required by the rule should be carried out in a concrete way, e.g., with the "method" (material technique) of laboratory analysis. Or else, to stay in the field of medicine, method in the second sense could include the rule that every hypothesis must be tested against public data, while in the third sense it might specify which instruments to use in the collection of data in order to respect this rule, e.g., the "method" (conceptual technique) of the double-blind.

15. The classical requisites for explication can be found in Carnap (1962), p. 3. For those indicated here, see Laudan (1983), p. 118. They correspond, respectively, to the requisites of similarity to the *explicandum,* and to Carnap's exactness. In (1987a), p. 20, Laudan claims that "the requirement that a methodology or epistemology must exhibit past science as rational is thoroughly wrong-headed." We shall return to the subject of the relationship between method and history, or the practice of science, in the next chapter. It is enough here to observe that a satisfying explication of the notion of method must save at least some eminent examples of scientific practice.

16. Feyerabend (1975), p. 295.

17. Galilei (1953), p. 356.

18. Ibid., p. 346.

19. Ibid., p. 345.

20. Ibid., p. 347.

21. Ibid., p. 352.

22. Compare this procedure with that indicated by C. Bernard: "the true scientist is one whose work includes both the experimental theory and experimental practice; (1)He notes a fact; (2) *a propos* of this fact, an idea is born in his mind; (3) in the light of this idea, he reasons, devises an experiment, imagines and bring to pass its material conditions; (4) from this experiment, new phenomena result which must be observed, and so on and so forth." Bernard (1865), p. 24.

This ordered sequence made up of four stages is the same procedure as that advocated by Beveridge in the following passage (in which point [a] figures as an external generator of the sequence rather than one of its stages): "the modern view of the scientific method may be summarized as: (a) recognition and formulation of the problem, (b) collection of relevant data, (c) arriving at a hypothesis by induction, indicating causal relations or significant patterns in the data, (d) making deductions from the hypothesis and testing the correctness of these by experimentation or collection of more data, (e) reasoning that if the results are consistent with the deduction, the hypothesis is strengthened but not proved." Beveridge (1980), p. 55.

As for Galileo, an analogous four-stage procedure can be found in his Paduan manual, *Treatise on the Sphere* (date uncertain). See Galilei (1605), pp. 211–12.

23. The idea that a change in theory also meant a change in method was clear to Bacon. See (1623), V, iii, p. 264: "let men be assured that the solid and true arts of invention grow and increase as inventions themselves increase." Peirce was of the same opinion. See (1877), 1: "each chief step in science has been a lesson in logic."

24. On the abandonment of the old *ars inveniendi* in favor of the hypothetico-deductive method, see Laudan (1981), chap. 11.

25. Risch (1980), chap. 1.

26. As regards analysis, see the following passage: "Progress in scientific work is just as it is in analysis. We bring expectations with us into the work, but they must be forcibly held back. By observation, now at one point and now at another, we come upon something new; but to begin with, the pieces do not fit together. We put forward conjectures, we construct hypotheses, which we withdraw if they are not confirmed, we need much patience and readiness for any eventuality, we renounce early convictions so as not to be led by them into overlooking unexpected

factors, and in the end our whole expenditure of effort is rewarded, the scattered findings fit themselves together, we get an insight into a whole section of mental events, we have completed our task and now we are free for the next one. In analysis, however, we have to do without the assistance afforded to research by experiment" (Freud 1933, p. 174).

As regards psychoanalytic theory, see the following declaration of principle: "Psychoanalysis is not, like philosophies, a system starting out from a few sharply defined basic concepts, seeking to grasp the whole universe with help of these and, once it is completed, having no room for fresh discoveries or better understanding. On the contrary, it keeps close to the facts in its field of study, seeks to solve the immediate problems of observation, gropes its way forward by the help of experience, is always incomplete and always ready to correct or modify its theories. There is no incongruity (any more than in the case of physics or chemistry) if its most general concepts lack clarity and if its postulates are provisional; it leaves the more precise definition to the results of the future work" (Freud 1922, pp. 253–54).

Jung's opinion on the method analytical psychology was to follow was similar. In one passage one can read: "A genuinely scientific attitude must be unprejudiced. The sole criterion for the validity of an hypothesis is whether or not it possesses an heuristic—i.e. explanatory—value. The question now is, can we regard the possibilities set forth above as a valid hypothesis? There is no *a priori* reason why it should not be just as possible that the unconscious tendencies have a goal beyond the human person, as that the unconscious can 'do nothing but wish'. Experience alone can decide which is the more suitable hypothesis" (Jung 1934, p. 131).

As for Freud, Grünbaum has quite rightly noted that "Freud's actual criteria for theory validation were essentially those of hypothetico-deductive inductivism" (Grünbaum 1986, p. 220). See also Grünbaum (1984), Introduction.

27. Federspil and Scandellari (1984), p. 440.

28. Bunge (1967), 1:40–41.

29. Eysenck (1953), pp. 228–29.

30. Legrenzi (1975), p. 53.

31. See Kant (1786), p. 7: "And since in every doctrine of nature only so much science proper is to be found as there is a priori cognition in it, a doctrine of nature will contain only so much science proper as there is applied mathematics in it."

32. This was Grünbaum's suggestion for Freud's etiology of paranoia. See, among other writings, Grünbaum (1984), chap. 1, sec. B, and Popper's answer in Grünbaum (1986).

33. Piaget (1965), p. 142.

34. It is not my intention to claim that introspection plays no role, or that its role should be relegated to a "context of discovery." I have criticized elsewhere the distinction between "context of discovery" and "context of justification" (Pera 1980), and I have maintained that any fact may be relevant in supporting an argument as long as there is an argumentative link between the two (Pera 1987). For a more detailed analysis of the role of introspection, see Barrotta (1992), p. 18.

35. See Eysenck (1953), pp. 232–35.

36. Ibid. p. 241.

37. An acceptance rule may be redundant and substituted by a preference rule if one considers that no piece of knowledge comes out of the blue but transforms previous knowledge. This issue will be examined later.

38. Galilei (1611), p. 142.

39. Ibid.

40. Galilei (1953), p. 115.

41. Ibid., p. 122.

42. Galilei (1615), p. 85.

43. Galilei (1624), p. 174.

44. Ibid., p. 174.

45. The conclusion Ingoli draws from what he considered definitive proof against Copernicus (which Galileo considered no more than an "anomaly") (Galilei 1624, p. 175) is: "From this it follows that the Copernican system is very doubtful, since it agrees in no way with the phenomenon it was constructed to save." See Ingoli (1616), p. 339.

46. Galilei (1953), p. 339.

47. This is my version of Feyerabend's view (1975), p. 23: "given any rule, however 'fundamental' or 'necessary' for science, there are always circumstances when it is advisable not only to ignore the rule, but to adopt its opposite." By accepting this view, I do not intend to say that whoever violates or ignores a fundamental rule behaves irrationally; he behaves irrationally only if rationality, as in the Cartesian project, is linked to the rules of method.

Two

1. In Laudan's words: "The fact that not all methods are permanent in science, the fact that new methodologies come to the fore, does not warrant the claim that all methodologies are equally good." See Laudan (1978), p. 536.

2. Though the methodology of Descartes was intentionally a priori since, as he said, its fruits come from "innate principles," this does not mean that he did not learn from these fruits. Indeed he was convinced, regarding his method, that "the great minds of the past were to some extent aware of it, guided to it even by nature alone" (1628, p. 17). Method, that is, was already at work in mathematicians; it was just a matter of discovering it.

It must also be recognized that even the so-called "Euclidean" or a priori methodologies were never such in absolute. As McMullin (1979) and Hempel (1983) quite rightly pointed out, not even Carnap could do away with scientific practice completely, especially because one of his requirements for an *explicatum* was that it should be "similar to the explicandum in such a way that, in most cases in which the explicandum has so far been used, the explicatum can be used" (Carnap 1962, p. 7).

3. Lakatos (1971), p. 122.

4. Ibid., p. 123.

5. R. Laudan et al. (1988), p. 5. See also L. Laudan et al. (1986). The attempt of Laudan's group to test methodologies with the history of science has been the most systematic yet.

6. Lakatos (1971), p. 124.

7. Laudan (1977), p. 160.

8. Lakatos (1971), pp. 103n and 117.

9. This is what I claimed in Pera (1986a), some of the themes of which are taken up here. After the "historicist turn," literature on the relationship between history and both the history of science and the history of philosophy has increased enormously. Regarding the various forms and ambiguities of historicism, McMullin (1979) is very useful.

10. Lakatos (1971), p. 124.

11. Laudan (1977), p. 158.

12. Ibid., p. 160.

13. See Lakatos (1971), p. 103n.

14. According to Goldberg (1984, p. 149), "the choice between the two theories was not made, ultimately, on the basis of the answers that they provided, but rather on the questions that they raised. The theory of relativity proved to be the heuristic theory." Resnick (1968, chap. 1.9), on the other hand, thinks it was rational to accept Einstein's theory because it not only succeeded in explaining known experimental results but it also predicted new effects.

15. Feyerabend (1976), p. 209.

16. For the whole story, see Goldberg (1984).

17. See the objections of Quinn (1971) and Musgrave (1976) to Lakatos.

18. I borrow the expression "open texture" from Waismann (1968), who used it to show the essential incompleteness of empirical concepts and descriptions.

19. Lakatos believed that his methodology was better than Popper's because it contains "fewer methodological decisions" (1970, p. 40), that is, fewer personal ingredients. This is a particularly important test for Lakatos because, as we have already said, he was affected by the Cartesian syndrome and thought that personal decisions that aren't regulated by appropriate rules are the Trojan horse within which irrationality is smuggled into science.

20. Lakatos (1971), p. 136.

21. Ibid., p. 137.

22. I examined Lakatos's project in greater detail in Pera (1989a).

23. This is Lakatos's autobiographical expression. See Lakatos (1978b), p. 61.

24. Scheffler (1967), pp. 9–10. Scheffler added a note of caution, but also a hint of trust: "We do not, surely, have explicit and general formulations of such criteria at the present time. But they are embodied clearly enough in scientific practice to enable communication and agreement in a wide variety of specific cases" (p. 10). A philosophically less sophisticated position than this, but following the same lines, can be found in the Eysenck passage quoted in the Introduction.

25. Laudan (1987b), p. 351.

26. See Laudan (1987a) and (1990), in which Laudan defends this point of view in the face of criticism by G. Doppelt, J. Leplin, and A. Rosenberg.

27. See Popper (1974), p. 1036.

28. Laudan (1987a), p. 24.

29. Ibid., p. 27.

30. Ibid.

31. Elsewhere Laudan has stressed that *logical* underdetermination does not imply *methodological* underdetermination. See Laudan (1990).

32. Laudan (1987a), p. 25.

33. Ibid.

34. Ibid., p. 26.

35. It is not certain that Laudan intends using a transcendental argument; he sometimes seems to accept (R1) either because it is a principle that "all the disputing theories of methodology share in common" (1987a, p. 25), or because it is "universally accepted by methodologists" (ibid., p. 29). This, however, raises well-known difficulties regarding the inductive justification of induction.

36. This version has been supported by Sklar (1983); see p. 153.

37. I consider them imperatives of prudence rather than of ability because, in scientific subjects, the goal of knowledge about the world is the equivalent of what Kant considers in practical subjects the "one end that can be presupposed as actual in all rational beings," that is, happiness. See Kant (1785), p. 83.

38. Kuhn (1962), p. 152.

39. Ibid., p. 94.

40. Rorty (1982), p. 165.

41. Ibid., p. 163.

42. *Topics* 100b21–23.

43. *Rhetoric,* 1354a1.

44. Ibid., 1356a25ff.

45. Ibid., 1355a5, 1356a35, 1400b35.

46. Ibid., 1356b5.

47. Ibid., 1355b15–17.

48. Perelman and Olbrechts-Tyteca (1958), p. 4.

49. See *Posterior Analytics,* 84a35ff: "for it is by interpolating a term inside and not by taking an additional one that what is demonstrated is demonstrated."

50. See the following reference made by Aristotle when the problem is raised in *Prior Analytics:* "thus we have explained fairly well in general terms how we must select propositions: we have discussed the matter precisely in the treatise concerning dialectics" (46a28–30).

51. *Topics,* 101a35ff. See also 163b9–10: "moreover, as contributing to knowledge and to philosophic wisdom, the power of discerning and holding in one view the results of either of two hypotheses is no mean instrument; for it then only remains to make a right choice of one of them."

52. *Rhetoric,* 1355a30ff.

53. On this double sense (*phainomena-legomena*) of "starting with details," see Owen (1961).

54. This explains the debate over his predecessors' views in many of Aristotle's scientific writings, from the *Physics* to *On the Heavens.* On the role of this debate, see Mansion (1961).

55. On this point, see Weil (1951). On the role of dialectics in science according to Aristotle, Berti (1987, 1989) is highly illuminating.

56. Galilei (1593), pp. 150–51.

57. See Galilei (1633), pp. 731ff. and the analysis of his note made by Pasquinelli (1968), pp. 175–79.

58. See Lakatos (1970), p. 46; the expression is Popper's (1945), 2:218.

59. See Perelman (1979), p. 138: "the system used by the judge . . . does not respond to the three exigencies of non-ambiguity, coherence and completeness."

60. See Perelman (1970), p. 143. It is interesting to note that immediately following this, Perelman adds that the same is not true for science.

61. See Perelman (1971), p. 154. On the lacunae in law, see texts collected by Perelman in Perelman, ed. (1968).

62. See Perelman, ed. (1965).

63. Hempel's most recent view is a partial remedy to this kind of ambiguity. As I shall attempt to clarify further on, my project is different from but not incompatible with Hempel's. I think my scientific dialectic catches the intentions, if not the content, of what he calls "relaxed methodology" (1983, p. 93).

64. The philosopher of law Uberto Scarpelli quite rightly noted that "if the structures to which the signs of law belong often have a vague and elastic character in most non-extreme cases, they are adequately and sufficiently determinant. Although an edge of uncertainty always seems to accompany the development of law, there is always a heart of certainty that allows it to work as a social structure". See Scarpelli (1982), p. 197.

65. Kuhn (1970a), p. 262.

66. I do not mean formal logic *applied* to law (e.g., Kalinowski 1965), but the logic *of* law (e.g. Perelman 1979 and Giuliani 1975).

Three

1. Feyerabend (1975), p. 152.

2. Ibid., pp. 153–54.

3. Galilei (1953), p. 268.

4. Ibid., p. 54.

5. The classification of arguments in this and in the following sections follows, where possible, Perelman and Olbrechts-Tyteca's (1958) terminology, among the richest available. Little indulgence, however, is required since the same rhetorical argument can reasonably be placed under different labels and can serve different functions.

6. Galilei (1953), p. 32.

7. The following passage, among many others, in which Aristotle argues against the Pythagoreans about the position of the earth, is a case in point: "they further construct another earth in opposition to ours to which they give the name counter-earth. In all this they are not seeking for theories and causes to account for the

phenomena, but rather forcing the phenomena and trying to accommodate them to certain theories and opinions of their own." See *On the Heavens,* 239a23–26.

8. Galilei (1953), pp. 55–56.

9. See ibid., p. 50 and p. 110.

10. Ibid., p. 14.

11. Ibid., p. 203.

12. Ibid., pp. 206 and 207.

13. Ibid., p. 207.

14. Galilei (1974), p. 12.

15. This is the same analogical argument used by Galileo in the famous thought experiment of the "large ship" (*gran naviglio*) to prove the principle of relativity. See Galilei (1953), pp. 186ff. This experiment and other similar ones as well are the basis of Feyerabend's analysis (1975, chap. 7) according to which, in order to persuade people to move on from the Aristotelian interpretation (every movement is operative) to the Copernican one (only relative movement is operative), Galileo had to resort to "arguments in appearance only," "propaganda," or "psychological tricks." Obviously Feyerabend has a restricted, Cartesian view of the nature of an argument. Galileo uses counter-examples and analogies; despite his claims (which are admittedly propagandistic), these are neither "sensory experiences" nor "necessary demonstrations," but they are not tricks either. Because they do not fit valid deductive or standard inductive molds, does not mean they are *illogical.* For an answer to Feyerabend along these lines, see Finocchiaro (1980), especially chap. 8.

16. The most far-reaching analysis of Galileo's rhetoric has been made by Finocchiaro (1980). See also Vickers (1983) and Moss (1983, 1984, and 1986). The ambiguity of the term "rhetoric" is reflected in these studies. Following Perelman's *New Rhetoric,* Finocchiaro mostly (but not exclusively) concentrates on the logic of Galilean discourse, while Vickers, following (part of) Aristotle's *Rhetoric,* focuses on modes of speech. Vickers's view that the *Dialogue* belongs to the genre of epideictic rhetoric is correct but incomplete. In epideictic rhetoric, the audience stays in silence. In the *Dialogue,* this is true for the relationship between the work and the audience to which it is directed, but it is not true of the relationship between the three protagonists, who participate actively in the discussion. Finocchiaro (1990) provides a useful map of the different meanings of rhetoric and their relationship to science.

17. Galilei (1953), pp. 33–34.

18. Ibid., p. 43.

19. Ibid., p. 59.

20. Ibid., p. 60.

21. Ibid., p. 271.

22. Ibid., p. 118.

23. Ibid., p. 120.

24. Ibid., pp. 270 71.

25. Ibid., p. 387.

26. Ibid., p. 389.

27. Ibid., pp. 399–400.

28. Ibid., p. 69.

29. Ibid., p. 107.

30. Ibid., p. 108.

31. Ibid., p. 109.

32. Ibid., p. 239.

33. Ibid., p. 410.

34. Galilei (1623), p. 269.

35. Galileo uses other expressions which express the same dilemma well. In his *Assayer,* he said that the book of nature was either like Euclid's *Elements* or like Homer's *Iliad.* See Galilei (1623), p. 237.

36. It was obviously not a great fortune for Galileo himself. Moss is right when she comments: "In the final analysis the problem the *Dialogue* created for Galileo was that his rhetoric worked too well." Moss (1984), p. 103.

37. See Darwin (1859), p. 459 and (1958), p. 140.

38. Darwin (1958), p. 119.

39. See his letter to Asa Grey dated November 20, 1859: "What you hint upon generally is very, very true: that my work will be grievously hypothetical, and large parts by no means worthy of being called induction, my commonest error being probably induction from too few facts" (Darwin 1903, 2:126).

40. See, in particular, Crombie (1960), Ghiselin (1969), Ruse (1971, 1975a, 1979), Caplan (1979). On Ghiselin's interpretation, see Egerton (1971).

41. See Thagard (1978).

42. Darwin (1859), p. 63.

43. Ibid., p. 467.

44. Ibid., p. 61.

45. Darwin (1872), p. 455. See also Darwin (1887), 17:437, and Darwin (1903), 1:455.

46. Darwin (1859), p. 481.

47. Darwin (1903), 1:139–40.

48. There is also a third kind of interpretation, based on the "semantic" approach to theory; see Lloyd (1983). Not satisfied with current interpretations, Hull (1973, p. 35) observed that "evolutionary theory and philosophy of science are still at odds," and Mayr (1991, p. 10) noted that Darwin's "procedure does not fit well into the classical prescriptions of the philosophy of science, because it consists of continually going back and forth between making observations, posing questions, establishing hypotheses or models, testing them by making further observations, and so forth."

49. Recker (1987) distinguished three strategies in the *Origin:* an "Empiricist Vera Causa Strategy" in chapters 1–4, a "Response to Objections Strategy" in chapters 6–9 and 11, and an "Explanatory Power Strategy" in chapters 10, 12, 13. Recker does not, however, say that the second strategy is mainly rhetorical. The term "rhetoric" seems so unpopular that it is not even adopted by Himmelfarb (1959) and Gale (1982) who interpret Darwin along these lines.

50. The correspondence between these sets of arguments and chapters in the *Origin* (first edition) is as follows: A = 1–4, B = 6–9, C = 10–13. This correspondence is only rough because some arguments from set B also belong to set C on the basis of the (rhetorical) principle that in a controversy an objection that has been overcome is equivalent to a proof in favor.

51. Newton (1726), 2:398.

52. See Whewell (1847), 2:286: "when two different classes of facts lead to the same hypothesis, we may hold it to be a *true* cause"; Herschel (1830), p. 149: "If the analogy of the two phenomena be very close and striking, while, at the same time, the cause of one is very obvious, it becomes scarcely possible to refuse to admit the action of the analogous cause in the other, though not so obvious in itself." Regarding Herschel and Whewell's influence on Darwin, see Ruse (1975b).

For more on the history of the principle of the *Vera Causa* and Darwin's role in it, see Kavaloski (1974) and Hodge (1977).

53. Darwin (1859), p. 61.

54. Ibid., p. 63.

55. Ibid.

56. Ibid., p. 61.

57. Ibid., pp. 80–81.

58. Ibid., p. 469.

59. Ibid., p. 61.

60. Ibid., pp. 83–84. For the narrative role of passages such as this and the previous one, see Richards (1992).

61. Darwin (1859), p. 186.

62. Ibid., p. 189.

63. Ibid., p. 280.

64. Ibid., p. 298.

65. Ibid., p. 188.

66. Ibid., p. 201.

67. Ibid.

68. Ibid., p. 167.

69. Ibid., p. 202.

70. Ibid., p. 237.

71. Ibid.

72. See note 39 above.

73. Sedgwick (1860), p. 160.

74. Hopkins (1860), p. 239.

75. Mill (1865), p. 308.

76. See note 46 above.

77. Darwin (1887), 17:108.

78. Darwin (1872), p. 81.

79. Ibid., p. 455. See also Darwin (1903), 1:184: "I am actually weary of telling people that I do not pretend to adduce direct evidence of one species changing into another. . . . I generally throw in their teeth the universally admitted theory of the undulations of light . . . admitted because the view explains so much."

80. Hopkins (1860), p. 231.

81. Darwin (1872), pp. 442–43.

82. Ibid., p. 227.

83. Ibid., p. 229.

84. Jenkin (1867), p. 311.

85. Darwin (1872), pp. 47–48.

86. Darwin (1859), p. 280.

87. Darwin (1872), p. 308.

88. Ibid., p. 90.

89. Ibid., Darwin (1872), p. 195; see also p. 22.

90. Ibid., pp. 90–91.

91. Darwin has often been criticized for this use of a "logic of possibility." See Whewell (1833, pp. xvii–xviii), Jenkin (1867, p. 339), and Himmelfarb (1959, p. 334). Harris (1970, pp. 186–87), on the other hand, considers this to be one of the secrets of the success of Darwin's theory, on the basis of the criterion that "when we see how phenomena fit into such a system we find them intelligible and accept the theory setting out the system as an adequate explanation". But this criterion does not come from any Moses-like methodology; it can only be introduced in a debate in which its legitimacy, interpretation, and restrictions are under discussion.

92. Darwin (1872), p. 190; my emphasis.

93. Jenkin (1867), pp. 338–39.

94. See Darlington (1959), p. 59.

95. See Himmelfarb (1959), p. 333.

96. See Gale (1982), p. 161.

97. Hoyle (1965), p. 81.

98. Weinberg (1977), pp. 8–9, 10.

99. Davies (1977), p. 141.

100. Davies also maintains that "science does not deal in beliefs, but in facts. A model of the universe does not require faith, but a telescope. If it is wrong, it is wrong" (1977, p. 201). But it is hard to believe that religious opinion has no role in modern cosmological matters, especially when one considers that one aspect of the controversy between the steady-state theory and the big-bang theory has to do with their different attitudes to the problem of "creation." According to Bondi (1960b, p. 140), with the steady-state theory "the problem of creation is brought within the scope of the physical inquiry." One of Hoyle's main reasons for being against the big-bang theory is that "it is against the spirit of scientific inquiry to regard observable effects as arising from causes unknown (and unknowable) to science, and this is what creation in the past implies" (Hoyle 1948, p. 372).

The relationship between cosmological theories and the problem of creation has been severely and convincingly criticized by Grünbaum (1989). Quite rightly, McMullin (1981a, p. 39) claims that "what one cannot say is, first, that the Christian doctrine of creation 'supports' the Big Bang model, or, second, that the Big Bang model 'supports' the Christian doctrine of creation." See also McMullin (1981b).

101. Bondi et al. (1960), p. 11.

102. Ibid. p. 36.

103. Ibid.

104. Ibid. pp. 13–14.

105. Ibid. p. 38.

106. Ibid., p. 45. See also Bondi and Gold (1948), p. 254.

107. Bondi et al. (1960), pp. 41–42.

108. Ibid. pp. 17–18.

109. Ibid., p. 42.

110. Ibid., pp. 48–49.

111. Ibid., pp. 42–43.

112. Ibid., pp. 45–46.

113. Ibid., p. 58.

114. Hoyle (1965), p. 81.

115. Ibid., p. 86.

116. See for example Bondi (1960b), pp. 5ff, where Bondi contrasts the "extrapolating" (that is inductive) with the "deductive" approach to cosmology and states that "the tendency on the part of the empirical school to regard any work of the deductive type as being 'too speculative' seems to be based on a complete misunderstanding of the scientific scale of values."

117. This technique appears to be typical. Laudan, who has dealt extensively with the problem of the choice of method, recalls how Lesage used to defend his use of the hypothetical method in physics: "His strategy was twofold: first, to establish that such a method promoted legitimate aims for science; second, to show that even his critics—in their actual practice—utilized unobservable entities" (Laudan 1984, p. 58). The second strategy is clearly based on an *ad hominem* argument.

118. Sciama (1973), p. 56.

119. Narlikar (1973), p. 69.

120. Bondi (1960a), p. 35.

Four

1. M. Salmon, for example, maintains that "arguments can be classified in terms of whether their premises provide (1) conclusive support, (2) partial support, or (3) only the appearance of support (that is no real support at all)." In the first case they are deductive, in the second inductive, and in the third fallacious (M. Salmon 1984, p. 32). Similarly, W. Salmon claims there are "two major types [of arguments]: deductive and inductive" and thus "there are logically correct forms of each, and we can roughly distinguish incorrect forms (fallacies) of each type" (W. Salmon 1984, p. 14). More precisely, having defined fallacies as "logically incorrect arguments," W. Salmon divides them into two groups: "invalid arguments," whose premises do not imply their conclusion logically, and inductive fallacies, that is, "incorrect arguments," whose premises do not support their conclusion (1984, pp. 17–18, 88). The question whether classifications of this kind are exhaustive has been discussed in Barker (1965). See also Hamblin (1970, chap. 7) and Wellman (1971).

2. Kant (1978), A61, B86.

3. Galilei (1953), p. 234.

4. It is still worth reading Aristotle. See *Prior Analytics,* 64b.28ff. "To beg and assume the point at issue [*petitio principi*] is a species of failure to demonstrate the problem proposed; but this happens in many ways. A man may not deduce at all, or he may argue from premises which are unknown or equally unknown, or he may establish what is prior by means of what is posterior; for demonstration proceeds from what is more convincing and prior. Now begging the point at issue [*petitio principi*] is none of these; but since some things are naturally known through themselves, and other things by means of something else (the first principles through themselves, what is subordinate to them through something else), whenever a man tries to prove by means of itself what is known of itself, then he begs the point at issue [*petitio principi*]." In other words, *petitio principi* does not violate the conditions of the syllogism but certain conditions of the demonstration (scientific syllogism); that is, it is fallacious not *qua* a syllogism but *qua* a demonstration.

5. See Woods and Walton (1982), pp. 96–97.

6. Here too it is worth reading Aristotle when he claims that a sophistical argument is "not only a deduction or refutation which appears to be valid but is not, but also one which, though it is valid, only appears to be appropriate to the things in question." See *Sophistical Refutations,* 169b20.

7. Quite rightly, T. Govier labeled logical dualism "positivist theory," "because of its relation to the positivistic theory of knowledge, according to which there are two and only two sources of genuine knowledge—the formal sciences and the empirical sciences." See Govier (1987), p. 57. An effective critique of logical dualism can be found in Finocchiaro (1980), chaps. 13 and 15.

8. See Polya (1954).

9. Ibid., p. 113.

10. Although De Finetti maintained that "inductive logic is reduced in essence

to the theorem of compound probabilities or to its slightly more elaborate variant, often called Bayes' theorem" (De Finetti 1972, p. 150), the logic of plausible reasoning does not cover everything that passes under the label of "inductive logic." For example, as W. Salmon has often stressed (1966, pp. 15–131; 1984, pp. 133ff; and 1991), Bayes's theorem explains one of the typical prescriptions of inductive logic (more precisely of the logic of confirmation), that H cannot be considered confirmed by O if H does not have a positive initial probability or if its initial probability is not greater than that of a rival hypothesis H'. However, the attribution of initial probability requires reasons. These reasons can be conferred through a stock of arguments, including analogy, simple enumeration, pragmatical arguments, arguments of the possible, etc. For these arguments, however, there is no logic that can be reduced to probability calculus.

11. See Barrotta (1992), 16, to whom I am indebted. Perelman's definitions of rhetoric are not univocal. At one time he writes that "what distinguishes rhetoric from formal logic or from positive science is that it still concerns itself with adherence more than with truth" (Perelman 1979, p. 107) and that "rhetoric tries to persuade through discourse. If in order to obtain adherence to an affirmation one relies on experience, this cannot be called rhetoric" (ibid., p. 105). At another time he puts deductive and inductive arguments under the label of "theoretical reasoning" and rhetorical arguments under that of "practical reasoning" and writes that "while theoretical reasoning consists of an inference that draws a conclusion from a proposition, practical reasoning is that which justifies a decision" (Perelman 1968a, p. 168), or that "practical reasoning has a different structure from theoretical reasoning which leads to the truth or to the probability of a conclusion or, at least, to the fact that can be correctly inferred from the premise" (Perelman 1970, pp. 184–85).

12. For this approach to fallacies, see Hamblin (1970), Woods and Walton (1982).

13. Concerning the relationship between formal logic and argumentation theory, see Krabbe (1982).

14. Finocchiaro used this example to raise the problem whether a presumed fallacy, based on the standard classification in logic textbooks, is actually a fallacy. See Finocchiaro (1980), p. 336, and (1981), p. 16.

15. See Aristotle's *Rhetoric,* 135a8ff: "the orator's demonstration is enthymeme, and this is, in general, the most effective of the modes of persuasion; the enthymeme is a sort of deduction (the consideration of deduction of all kinds, without distinction, is the business of dialectic, either of dialectic as a whole or one of its branches): clearly then, he who is best able to see how and from what elements a deduction is produced will also be best skilled in the enthymeme when he has further learnt what its subject-matter is and in what respects it differs from the deductions of logic."

16. See Perelman and Olbrechts-Tyteca (1958), part 2.

17. A few formal systems of dialogue-games already exist. See, for example, Barth and Krabbe (1982) and Krabbe (1978). Their application to scientific dia-

logues has still to be explored. Hintikka's "interrogative model" (see, for example, Hintikka 1987, 1988, 1989, and Hintikka and Hintikka 1982) is a dialogue-game against *nature.* In the terms to be introduced in the next chapter, it is still a game with two players.

18. "To come to another point, I entirely agree with you that general relativity is an excellent theory for the description of local phenomenon, like motions in the solar system. But do you find it useful in the extrapolation of the universe?" See Bondi, et al. (1960), p. 46.

19. Kant (1790), p. 21.

20. This was Planck's assumption: "Of course, it may be said that the law of causality is only after all a hypothesis. If it be a hypothesis, it is not a hypothesis like all the others, but it is a fundamental hypothesis because it is the postulate which is necessary to give sense and meaning to the application of all hypotheses in scientific research." See Planck (1932), p. 150.

21. Bondi and Gold (1948), p. 255 (1.2).

22. Bondi et al. (1960), p. 48.

23. Galilei (1953), p. 34.

24. Galileo's *Dialogue* is revealing on this point. Simplicio often objects to Salviati's "sensory experience," but he never doubts that cognitive claims must agree with empirical evidence.

25. This problem can be illustrated with the following exchange of remarks between Bonnor and Bondi: "*Bonnor:* I should like to repeat my point about simplicity. It is obviously simpler to postulate that energy is precisely conserved, and I am hoping that somebody will explain what dividend we can expect from making the principle of energy more complicated. *Bondi:* I believe that we get the dividend in the resulting picture of the structure of the universe. An unchanging universe is simpler than the evolving type of universe that you favor" (Bondi et al. 1960, p. 43). On Bondi's concept of simplicity, see also Bondi (1960b), pp. 13–14.

26. An illustration of this second problem is provided by the following objection from Simplicio: "It seems to me that you base your case throughout upon the greater ease and simplicity of producing the same effects. As to their causation, you consider the moving of the earth alone equal to the moving of all the rest of the universe except the earth, while from the standpoint of action, you consider the former much easier than the latter. To this I answer that it seems that way to me also when I consider my own powers, which are not finite merely, but very feeble. But with respect to the power of the Mover, which is infinite, it is just as easy to move the universe as the earth, or for that matter a straw. And when the power is infinite, why should a great part of it be exercised rather than a small? From this it appears to me that the general argument is ineffective." See Galilei (1953), pp. 122–23.

27. The following passage from Bondi illustrates this point clearly: "In particular, the tendency on the part of the empirical school to regard any work of the deductive type as being 'too speculative' seems to be based on a complete misunderstanding of the scientific scale of values. The overriding principle must be that of

the economy of the hypotheses. . . . The value of a hypothesis depends primarily on its fruitfulness, i.e. on the number and significance of the deductions that can be made from it, and not on whether it requires a change in outlook and is considered 'upsetting'." See Bondi (1960b), pp. 5–6.

28. Perelman and Olbrechts-Tyteca (1958), p. 84.

29. For the commonplace of fruitfulness, see Bondi et al. (1960), p. 21; Bondi and Gold (1948), p. 256 (2.3). For intertheory consistency, see Bondi et al. (1960), p. 45. The use of simplicity as a commonplace is frequent in Galileo. See Galilei (1953), pp. 118, 122, 327, 341, 344.

30. Although several authors take assumptions as presumptions rather than descriptive assertions (Toulmin 1953, pp. 144–48; Rescher 1977, pp. 113–16), it would seem preferable to consider them as the ontological foundation of presumptions. Thus, the assumption that nature is simple is the foundation of the presumption that "If nature is simple, then *simplex sigillum veri*." Assumptions cover and lay the foundation for presumptions.

31. Laudan (1984), p. 63.

32. I agree with Kuhn when he writes that "the criteria of choice . . . function not as rules, which determine choice, but as values, which influence it." See Kuhn (1977), p. 331.

33. This explication adapts one advanced by Lorenzen (1961, p. 195) later taken up by Barth and Krabbe (1982, p. 54). Lorenzen's definition is: "a step from a set, P, of premisses to a conclusion, Z, is dialectically valid (in a system s) if and only if: there is (given the dialectic system s) a winning strategy for a Proponent of Z, relative to P as the set of concessions made by the Opposition (in a discussion carried out according to the rules of the system s)." This definition represents the "dialectical garb of logic" as understood by Barth and Krabbe (1982) and Krabbe (1982).

34. It may be observed that, since it makes reference to substantive and procedural factors that can be specified, this explication makes a definition such as Wellman's (1971, p. 90) — "to say that an argument is valid is to claim that when subjected to an indefinite amount of criticism it is persuasive for everyone who thinks in the normal way" — more precise.

35. See Perelman (1977), p. 140: "two qualities are mixed together in such a way that they are hard to keep apart. These two qualities are *efficacy* and *validity*."

36. Perelman and Olbrechts-Tyteca (1958), p. 465. See also Perelman (1970), pp. 53, 62.

37. Concerning these difficulties, see Apostel (1979).

38. See Perelman (1977), p. 140.

39. Ibid.

40. I have adapted the definition given by Barth and Krabbe (1982), p. 83, to my own ends.

41. For a formal system of dialectics, see Barth and Krabbe (1982), chap. 4.5.

42. Rescher (1977) has a more sophisticated treatment of typical dialectical moves and counter-moves.

43. *Sophistical Refutations*, 164b3.

44. I am indebted to W. A. de Pater for this diagram which he uses to illustrate the role of Aristotle's *topics*. See de Pater (1965), p. 148.

45. Aristotle's definition of *topos* (the only one in his texts) is given in *Rhetoric* 1403a16–17: "an element (*stoikeion*) is a commonplace (*topos*) embracing (*eis . . . empiptei*, under which are included) a large number of particular kinds of enthymeme."

46. It is to be supposed that this is what Aristotle meant when he wrote (see previous note) that *topoi* "embrace" enthymemes.

47. Galilei (1953), p. 11.

48. The same rule was at work in the debate between Einstein and supporters of the "orthodox" interpretation of quantum mechanics. Einstein lost the debate for various reasons, in particular because he did not succeed in offering something more than an act of faith in support of his view. The following frank admission of defeat contained in a letter to Born dated March 3, 1947, illustrates (C1) very well: "But I am convinced that someone will eventually come up with a theory whose objects, connected by laws, are not probabilities but considered facts, as used to be taken for granted until quite recently. I cannot, however, base this conviction on logical reasons, but can only produce my little finger as witness, that is, I offer no authority which would be able to command any kind of respect outside my hand" (Einstein and Born 1971, p. 158). See also p. 180: "I have more or less understood your theoretical hints. But our respective hobby-horses have irretrievably run off in different directions—yours, however, enjoys far greater popularity as a result of its remarkable practical success, while mine, on the other hand, smacks of quixotism, and even I myself cannot adhere to it with absolute confidence. But at least mine does not represent a blind-man's buff with the idea of reality. My whole instinct rebels against it."

49. Rescher (1977), p. 27, distinguishes the two cases.

50. Bondi et al. (1960), p. 38.

51. Bondi and Gold (1948), p. 255.

52. Ibid: "we do not claim that this principle must be true."

53. Galilei (1953), p. 108.

54. Ibid., p. 234.

55. I here use terminology introduced by Laudan (1984).

56. Galilei (1953), p. 341.

57. Carnap defines "Ri⊣ Rj" as follows: "at least one of Ri's utterances is not true and at least one of Rj's is true." See Carnap (1959), p. 152.

58. Feyerabend (1981), 2:148.

59. Hempel (1983), p. 93.

60. Ibid.

61. That methodology is a "supplement" of deductive logic is Popper's thesis. See Popper (1959), p. 54. Neopositivists would have considered it a supplement of inductive logic. Quite rightly, Hempel has written that "the problem of formulating norms for the critical appraisal of theories may be regarded as a modern outgrowth of the classical problem of induction." See Hempel (1983), p. 92.

62. On the question of relations between the theory of argumentation and logic, see Krabbe (1982).

Five

1. Bacon (1620a), p. 39 (Preface).

2. Gassendi (1658a), p. 79b.

3. Ibid. (1658b), p. 207.

4. Bacon (1620a), p. 42 (Preface).

5. Ibid., p. 52 (Book 1, #29); see also (1620b), p. 24.

6. Bacon (1607), p. 601.

7. Galileo never considered Aristotle a corruptor, and as regards Aristotelian logic, he was not so much against dialectics per se as against the sophistic, degenerate use made of it by its followers at the time. Barone has written an excellent essay on Galileo's logic, showing how his criticism of syllogistic reasoning (or formal logic) was directed against precisely those "altercations" resulting from its misuse. See Barone (1967).

8. This accusation is frequently made in the *Assayer* against Father Grassi.

9. Galileo (1610), p. 423. At other times he said "like the *Iliad* or *Orlando Furioso.*" See Galileo (1623), p. 232.

10. Descartes (1628), p. 37 (Rule X).

11. Hooke (1665), Preface.

12. Sprat (1667), p. 62.

13. Hobbes (1642), p. 154.

14. Locke (1692), p. 283 (Book IV.7.11).

15. Rorty is among those who have forgotten this. See Rorty (1989), p. 6: "Europe did not *decide* to accept the idiom of Romantic poetry, or of socialist politics, or of Galilean mechanics. That sort of shift was no more an act of will than it was the result of argument. Rather, Europe gradually lost the habit of using certain words and gradually acquired the habit of using others." I think it would be impossible to do less justice to so many people and so many events in only one sentence.

Feyerabend is guilty of similar injustice. Comparing entities postulated by science with Greek gods, he has written that "the Greek gods were not 'refuted by argument'" and that "history, not argument, undermined the gods" (Feyerabend 1990, p. 144–45). I think that here, as elsewhere, "argument" is taken in a very restricted sense: is not persuasion that leads to conversion (or to put it in Aristotelian terms, prudence that causes a deliberation) a legitimate form of argumentation?

16. Bacon (1607), p. 601.

17. Descartes (1628), p. 15.

18. Ibid., p. 16.

19. Matthew, 5:37.

20. See Goethe (1949), p. 510. Goethe's aphorism was first used by Hermann Weyl and taken over by Popper (1959, p. 311). Lakatos corrected it, claiming that

"nature may shout NO, but human ingenuity—contrary to Weyl and Popper—may always be able to shout louder" (Lakatos 1971, pp. 150–51). His "sophistications" of Popper's methodology were his attempt to silence or at least hush the devil's voice ("mob psychology" in his terminology).

21. See, for example, *Posterior Analytics* 81a38–39: "it is evident that if some perception is wanting, it is necessary for some understanding to be wanting too."

22. Along with the passage indicated in chapter 3.8, see the following passage from *On the Heavens,* 306a5–17, in which Aristotle criticizes the theories of certain Platonists on generation: "It is absurd, because it is unreasonable that one element alone should have no part in the transformations, and also contrary to the observed data of sense, according to which all alike change into one another. In fact their explanation of the phenomena is not consistent with the phenomena. . . . But they, owing to their love for their principles, fall into the attitude of men who undertake the defence of a position in argument. In the confidence that the principles are true they are ready to accept any consequence of their application. As though some principles did not require to be judged from their results, and particularly from their final issue. And that issue, which in the case of productive knowledge is the product, in the knowledge of nature is the phenomenon always and properly given by perception." See also this passage from *On Generation and Corruption,* 316a5–10, in which Aristotle criticizes followers of Plato and Democrites: "lack of experience diminishes our power of taking a comprehensive view of the admitted facts. Hence those who dwell in intimate association with nature and its phenomena are more able to lay down principles such as to admit of a wide and coherent development; while those whom devotion to abstract discussions has rendered unobservant of the facts are too ready to dogmatize on the basis of a few observations."

23. See the following passage from *On the Heavens* in which Aristotle refutes Thales and his followers' opinion that the Earth floats on water: "These thinkers seem to push their inquiries some way into the problem, but not so far as they might. It is what we are all inclined to do, to direct our inquiry not to the matter itself, but to the views of our opponents; for even when inquiring on one's own one pushes the inquiry only to the point at which one can no longer offer any opposition. Hence a good inquirer will be one who is ready bringing forward the objections proper to the genus, and that he will be when he has gained an understanding of all the differences" (294b 6–13).

24. Kuhn (1977), p. 110.

25. As we have already mentioned (chapter 4), this is how methodology is conceived in Popper (1959), p. 54.

26. See Carnap (1962), no. 44. Carnap wrote (p. 203) that "the methodology of induction gives advice how best to apply the methods of inductive logic for certain purposes. We may, for instance, wish to test a given hypothesis *h;* methodology tells us which kinds of experiments will be useful for this purpose by yielding observational data e_2 which, if added to our previous knowledge e_1, will be inductively highly relevant for our hypothesis *h,* that is, such that $c\ (h, e_1 \wedge e_2)$ is either considerably higher or considerably lower than $c\ (h, e_1)$."

27. This is Feyerabend's expression based on the view that "empiricism insofar as it goes beyond the invitation not to forget considering observations, is . . . an unreasonable doctrine." See Feyerabend (1969), pp. 134–35.

28. Rescher observes quite rightly (1977), p. 112: "the dialectical model does not dismiss the standard evidential considerations or deny them the pivotal force that is their due. Quite to the contrary, these now appear as playing a vital role—but one that is played out *within* the framework of the dialectical process."

29. Doppelt claims that from Kuhn's view there follows a "moderate relativism concerning scientific rationality" according to which "scientific development is a rational process, in which the basic scientific choices are reasonable, but underdetermined by reasons and thus in need of an *ineliminable* dimension of sociological and psychological explanation" (Doppelt 1983, p. 111). The dialectical model upheld here does not intend to eliminate this dimension but to make it an *internal* factor that operates through the filter of scientific debate. In this perspective, the dominant cultural ideas of a society determine a theory choice only when they are transformed and become an integral part of the basis of scientific dialectics. For example, the Renaissance sun cult became an epistemic value (simplicity and harmony) or an assumption (the regularity of movement around the most noble body) that supported Copernicus' theory. Only a debate, following factors proper to it, can ultimately determine theory choice.

30. Bloor (1976), p. 32, establishes an analogy between the ways in which experience and social variables influence the acceptance of a cognitive claim and the parallelogram of forces, and, in the second edition (1991, p. 166), he stresses that knowledge does not depend *exclusively* on social variables. But knowledge is not a mediated response to a simple combination of forces; per se this combination is banal. Knowledge is the *direct* result of a debate conducted in terms of the factors of scientific dialectics in which many forces (that cannot be distinguished as either internal or external) are blended together.

31. Feyerabend (1970), p. 229.

32. See Rorty (1985), p. 6, for whom relativism "is the view that there is nothing to be said about either truth or rationality apart from descriptions of the familiar procedures of justification which a given society—*ours*—uses in one or another areas of inquiry." This is also the way relativism is viewed by many sociologists of science. See Barnes and Bloor (1982).

33. See Feyerabend (1978), p. 83, in which he defends that kind of relativism according to which "'Aristotle is true' is a judgement that presupposes a certain tradition, it is a relational judgement that *may* change when the underlying tradition is changed."

34. This is Wittgenstein's position (1953), although it is not clear whether he was only a virtual or an actual relativist. Marconi (1987), chap. 7, is illuminating on this point.

35. See Barnes and Bloor (1982), p. 27.

36. See Hacking (1982), p. 64.

37. This is the position maintained by Putnam, who defined it as "internal real-

ism." This holds that "a sign that is actually employed in a particular way by a particular community of users can correspond to particular objects *within the conceptual scheme of those users.* 'Objects' do not exist independently of a conceptual scheme." See Putnam (1981), p. 52.

38. See Winch (1958), p. 15: "our idea of what belongs to the realm of reality is given for us in the language that we use." This is also known as the Sapir-Whorf hypothesis. See Whorf (1954), p. 213: "the world is presented in a kaleidoscopic flux of impressions which has to be organized by our minds—and this means largely by the linguistic system in our minds." Sapir (1929), p. 209: "The worlds in which different societies live are *distinct* worlds, not merely the same world with different labels attached."

39. This is the type of relativism at the root of Kuhn and Feyerabend's incommensurability thesis. See Kuhn's well-known thesis (1962), p. 118, that different scientists with different paradigms work in different worlds.

40. For a partially similar list, see the Introduction to Hollis and Lukes (1982).

41. This is what it seems one can derive from the following version of Feyerabend's relativism: "For every statement, theory, point of view believed (to be true) with good reasons there exist arguments showing a conflicting alternative to be as good, or even better" (Feyerabend 1987, p. 76). Notice that the arguments referred to here must be *internal* to each theory, or else there would be common meta-arguments and meta-criteria for the two rival theories, contrary to the relativist view. However, if the arguments are internal to each theory, it is hard to understand how they can show that one theory is "at least as good, or even better" than its rival. The result, more likely, is that any theory is as good as another.

42. See the quotation from Rorty in note 32 above. See also Barnes and Bloor (1982), p. 34: "for the relativist there is no sense to be attached to the idea that some standards or beliefs are really rational as distinct from merely locally accepted as such."

43. See Rorty (1980), p. 331.

44. See Rorty (1989), p. 8.

45. Ibid., p. 6. Immediately after this, there follows the passage quoted in note 15 which states that "Europe did not *decide,*" etc. In line with thesis (c), Rorty here replaces an argued choice with an acritical habit: "that sort of shift was no more an act of will than it was a result of argument. Rather, Europe gradually lost the habit of using certain words and gradually acquired the habit of using others." It remains a mystery how habits of this kind can be acquired if not as the "result of argument."

46. Feyerabend (1987), p. 80. In this passage, the emphasis must be placed on "power-play" rather than on the possible "arguments," because, for the reasons given in note 41 above, these arguments are necessarily circular. Consistently enough, Feyerabend defined argumentative standards as "verbal ornaments" and spoke of recourse to "means other than arguments." See the passages quoted in the Introduction.

47. Defining tradition is complicated and possibly useless. But establishing if two communities, cultures, ways of thinking, theories, etc. belong to the same

tradition is perhaps easier. If one analyzes the polemic between, say, Galileo and Aristotle, or certain peripatetics such as Antonio Rocco, Ludovico delle Colombe, Francesco Ingoli, Giulio Cesare La Galla, etc. one finds that the sides diverge on a series of very important points but also agree on others, for example, on the idea that observation plays a central role in understanding nature. Together with others, this idea constitutes the nucleus of a minimum common framework that establishes contact between the two sides. Starting with that nucleus, argumentation is possible. When it becomes difficult, there is always the possibility of arguing *ad hominem* (a technique in which Galileo was supreme); however this equally presupposes a common idea, because *ad hominem* argumentation is based on the presumption of the value of noncontradiction. On methodological conservatism, see Barrotta (1992), pp. 141–43.

48. A possible objection to the dialectical model crops up here. This model seems to be conservative because the configuration of factors accepted in a given situation naturally, one could say "inertially," favors those theses whose support arguments are compatible with it and raises barriers against new ideas. But this is not necessarily the case. One must distinguish here between the *logical condition of argumentation* and a *heuristic suggestion for research*. During a debate, the accepted configuration may change. If in the beginning of the debate two interlocutors are committed to different configurations and then, in the course of the debate, they find a minimum configuration on the grounds of which to exchange views, then they can change their initial configurations. Thus, it can happen that a new thesis initially out of favor with the configuration in force can succeed in taking over if it is well defended. The dialectical model does not recommend conservatism. It simply recommends preferring the theses supported by stronger arguments.

The strategy for finding a minimum configuration is presumably the same as the one Quine defines as "semantic ascent" (Quine 1960, 56). At least in the dialectical model, this strategy explains—but does not recommend—what Quine has called "our natural tendency to disturb the total system as little as possible." See Quine (1953), p. 44.

49. Popper (1987), p. 36.

50. Ibid., p. 51.

51. Popper (1959), p. 37; my emphasis. Popper added a new note to the original passage quoted here (n. 5, p. 18) in which he wrote that "a reasonable discussion is always possible between parties interested in truth, and ready to pay attention to each other." A similar concept can be found in chapter 24 of Popper (1945).

52. Rorty (1982), p. 193.

53. Ibid.

54. Doppelt claims that "the good reasons in favor of a new paradigm are never more rationally compelling to scientists than those in favor of its predecessor unless it is already a process of conversion *to* the new standards implicit in it. The good reasons favoring the new paradigm provide *boundary conditions* on such 'conversions' preventing them from being irrational or unscientific" (Doppelt 1983, pp. 113–14). In the position upheld here strong arguments *cause* conversions. This does

not mean that a scientist cannot be converted to a new theory before being in possession of strong arguments; it means rather that the conversion is not rational until it has the support of strong arguments. What comes before the arguments only counts as an inclination, a psychological urge to look for supporting reasons.

55. Ronchi (1978), p. 853.

56. Feyerabend (1975), p. 81.

57. Rorty (1980), p. 331.

58. I have examined the exchanges of arguments and counter-arguments on this matter in more detail in Pera (1991b).

59. Scheiner (1612a), p. 25.

60. Ibid. p. 30.

61. Ibid.

62. Shea (1972), p. 51.

63. Ibid., p. 53.

64. Galileo (1663), p. 139.

65. See, however, the passage quoted above, in chapter 3, n. 8, and in n. 22 of this chapter.

66. Galileo (1613), p. 232n.

67. Note that Galileo's arguments against the assumption of incorruptibility were all the more forceful because Scheiner honestly recognized that the Aristotelian-Ptolemaic system, of which this assumption was an essential part, had to be rejected. At the end of his *Accuratior Disquisitio,* he wrote: "nevertheless . . . whether the spots are on the Sun or around the Sun, whether we say they are generative or not, whether we call all these things that vacillate clouds or not, what follows seems certain according to the common opinion of astronomers: that the density and constitution of the heavens as we consider it today can no longer be maintained, especially as far as the sun and Jupiter are concerned. We must therefore listen to the leading mathematician of our day, Christopher Clavius, who has warned astronomers in his latest book to provide a new system for the heavens because of these new—though very ancient—phenomena which can hardly be observed" (Scheiner 1612b, pp. 68–69).

68. Galileo's argument against the assumption of the incorruptibility of the heavens, expressed in terms of dialectical implication introduced in the previous chapter, is: $\{I, S\} \dashv \{T_1, T_2, T_3\}$. The decreased credibility of this assumption can be reconstructed following the pattern of Polya's logic of plausible reasoning (where | means "is incompatible with"):

$$\frac{\begin{array}{l}(I,S) \mid (T_1, T_2, T_3)\\ (T_1, T_2, T_3) \text{ is more credible (conceded by interlocutor)}\end{array}}{(I,S) \text{ is less credible.}}$$

69. See the quotations in n. 11 in chapter 4 above. As regards the distinction between rhetoric and science, Perelman's reaction to M.Polanyi's ideas is illuminating: "in my view, he puts the commitment of a scientist too much on a level with that of an artist or a philosopher. I think that there are essential differences between these subjects. Like Polanyi, I too defend personal knowledge as part of a culture and a tradition, but I would stress more than he does the absolutely particular

position the sciences occupy in our society, because in the sciences testing and verification techniques basically allow an agreement to be reached which is not possible in other domains" (Perelman 1970,p. 352).

70. Poznanski (1968), p. 72.

71. Scarpelli (1986), pp. 11–12.

72. See Perelman (1963), p. 93: "the logic of discovery obeys neither the formal schemes of deductions or the canons of Mill." See also Perelman (1968b), p. 508.

73. See Perelman (1970), p. 349.

74. See Perelman (1963), p. 92.

75. Walton builds several bridges between rhetoric and truth. He maintains, for example, that "positional argumentation is valuable and appropriate when the truth is not to be had by any other means." See Walton (1985), p. 263.

76. See also Popper (1959), p. 109: "what ultimately decides the fate of a theory is the result of the test."

77. See Preti (1968), pp. 199–200.

78. "Realism about theories" is an expression introduced by Cartwright (1983) and Hacking (1983), to distinguish it from "realism about entities." But realism has many varieties. For a list, see Horwich (1982) and Niiniluoto (1987a). The central theses of realism are listed in Boyd (1984), Leplin (1984), Niiniluoto (1987b).

79. *Phaedrus,* 260c.

80. This is the view of Putnam, among others. See (1981), p. 49: "'Truth' in an internalist view, is some sort of (idealized) rational acceptability—some sort of ideal coherence of our beliefs with each other and with our experience as those experiences are themselves represented in our belief system." See also p. 55: "truth is an *idealization* of rational acceptability."

81. For this and other criticisms, see Horwich (1990), p. 63.

82. On this point see Fine (1986), p. 141.

83. Popper (1983), p. 25.

84. Popper (1972), p. 103.

85. Putnam (1983), 3: xvii.

86. Kant would have agreed. As is well known, he defined truth as "the agreement of knowledge with its objects" (Kant 1978 A 58, B 82. See also A 157, B 197; A 191, B 236; A 642, B 670; A 820, B 848; Kant 1968b, p. 51; Kant 1988, Introduction, VII, p. 56) But this is a nominal definition of truth. Kant writes that "the truth depends upon agreement with the object, and in respect of it the judgements of each and every understanding must therefore be in agreement with each other (*consequentia uni tertio, consentiunt inter se*)" (A 820, B 848). Thus an agreement with the object exists when there is agreement with the facts. But "I cannot assert anything, that is, declare it to be a judgement necessarily valid for everyone, save as it gives rise to conviction" (A 821, B 849). Since conviction is that belief that "is valid for everyone, provided only he is in possession of reason" (A 820, B 848), it follows that, according to Kant, we have the right to consider a

cognitive claim true when there is general agreement over it. If this consensus exists, "there is then at least a *presumption* that the ground of agreement of all judgements with each other, notwithstanding the differing characters of individuals, rests upon the common ground, namely upon the object, and that it is for this reason that they are all in agreement with the object—the truth of the judgement being thereby proved" (A820–21, B 848–49).

87. See Rorty, respectively (1989), p. 8 and (1985), p. 6.

88. Watkins (1984), p. 126.

89. Ibid., p. 126.

Six

1. See Lakatos (1970), p. 46; Lakatos's expression is taken from Popper (1945), 2:218.

2. About this use, see Pera (1982).

3. Feyerabend (1962), p. 45.

4. Hanson (1969), 13.

5. Feyerabend (1965a), p. 214.

6. Hanson (1958), p. 36.

7. Despite the differences between the position supported here and Kant's, I would like to underscore that one can find three different theoretical levels in Kant. These are: (1) the forms of intuition of space and time and the categories; (2) those regulative ideas of reason that have an empirical use, such as "pure earth," "pure water," etc. (Kant 1978, A 646, B 674), as well as those principles or presuppositions, such as the ideas of homogeneity, continuity, etc. that "possess, as synthetic *a priori* propositions, objective but indeterminate validity, and serve as rules for possible experience" (Kant 1978, A 663, B 691); (3) empirical explicative theories, like Newton's, Copernicus' etc. These three levels correspond, in our language, to C-, I-, and E-theories.

8. Reference is made, in particular, to Lorenz (1941), von Bertalanffy (1955), and Popper (1972).

9. Useful information regarding this matter may perhaps be obtained from studies on "naive physics." See, for example, McCloskey (1983) and Bozzi (1990).

10. According to Heisenberg, this circumstance is at the origin of the paradox of quantum mechanics, because we are obliged to use concepts taken from classical mechanics in order to describe experimental results that at the same time cannot be explained in terms of those concepts (see Heisenberg, 1958, p. 45). This does not imply, however, that the categories are the same as the fundamental concepts of classical mechanics.

11. The distinction between these two types of seeing is taken from Dretske (1969). See also Brown (1987).

12. Regarding theoretical entities, see Shapere (1982) and Nola (1986).

13. A similar position is maintained by Brown (1977), p. 93: "we shape our precepts out of an already structured but still malleable material. This perceptual material, whatever it may be, will serve to limit the class of possible constructs without

dictating a unique precept." Note, however, that in the view maintained here data limit the range of concepts (predicates) that can be used to interpret them, not because of themselves but because certain concepts would not withstand a debate. To hold that data set limits by themselves is to go back to the empiricist view of a pure and absolute foundation of knowledge. A hint of this view seems to emerge from the following passage in Brown (1987), p. 199: "Thus, while the advocates of theory-ladenness are correct in emphasizing that a given body of data may be compatible with a large number of quite different beliefs, many of those who hold this view have ignored the equally important point that a given body of data will be *incompatible with an enormous number of beliefs*. It is this last point that captures the way in which the physical world constrains our observations, and it is through this constraint that observation provides the foundations for objectivity."

In one of Brown's earlier works, there was no such reference to foundations; however, he had written: "Unlike the Kantian position, or, rather, one interpretation of the Kantian position [that is, Rorty's], I do not maintain that theories impose structure on a *neutral* material" (Brown 1977, p. 93). We can see that the distinction between different theoretical levels turns out to be quite useful here. According to Kant, while categories (C-theories) impose a structure on neutral material, E- and I-theories impose a structure only on what Brown calls "already structured material." Rorty (1982, chap. 1) follows Davidson (1974) when he casts doubt on "the very idea of a conceptual theme" because he interprets Kant's constitution of the object as a twofold process: first (both in terms of time and logic) data, then schemes. But in Kant's view, this process is single: data and schemes (categories, but *not* theories) are one and the same thing.

14. Radder (1988, p. 105) has stressed this point very effectively: "the relation between the theoretical description and the material realization is in my opinion of a *causal* nature, in the sense that there is a material reality which is (in part) the cause of our theoretical terms referring. Those philosophers who try to translate the theoretical description into a *description* of the material realization attempt the impossible transformation of a causal relationship into an *epistemological* one."

15. This way of defining a predicate is taken from Agazzi (1976).

16. In order to define *B*, Agazzi omits *T* and uses the quadruple <*S*, *O*, *R*, *P*> = *B*. This is inadequate as a definition of the meaning of *B*. To take Agazzi's own example, suppose an individual is given the following instructions: "take the instrument *S* (an electroscope), carry out operations *O* (put *x* in contact with *S*'s plate), check result *R* (*S*'s gold leaves divaricate)." In this way, our individual will certainly learn the *name* "electrically charged" but will still be ignorant of its *meaning*. He will know the meaning of *B* only when he understands why he obtains *R* by carrying out *O* using *S*, that is to say, when he understands that by doing this a fluid is condensed, or a movement of particles is produced, etc.; that is, when he has accepted a theory. If I am not mistaken, Radder's criterion of reference (1988, p. 102) is the same as Agazzi's.

17. Agazzi writes that "the fact that in experimental science there are propositions which turn out to be forbidden, because certain conditions of referentiality (experimental results) are opposed to them is already an important symptom of the

fact that these propositions speak of reality" (Agazzi 1989, p. 88). In the view upheld here, this fact *guarantees* that such propositions have a *putative* reference, but it is not a symptom that they have a *real* reference.

18. In the Italian edition of this book I used the expression "reference without reality." I later discovered it had been introduced by Gross (1990). I have therefore changed the expression slightly, by no means in order to set myself apart from Gross with whom I agree on this point, but in order to avoid idealistic misunderstandings and because, as I explain in the text, I think that a deflationary strategy allows us to keep the notion of reality without counter-intuitive consequences.

19. Franklin (1941), p. 174.

20. I think this view is pretty close to that of Baltas. He writes: "the object of Physics is constructed by Physics' conceptual system itself, and internally structured by the experimental procedures specific to that science" (Baltas 1987, p. 135); for example, "the *natural* phenomenon of a falling apple is transformed by the conceptual system of Classical Mechanics into the *physical* phenomenon of a mass point attracted by a gravitational force" (p. 133). Baltas also quite rightly stresses the role that "ideological assumptions" (something like, I suspect, what I call I-theories), play in this transformation. As he writes, "physical phenomena do not just lie there, ready to be explained by some theory which is devised to that end by some physicist's cleverness or genius. These are phenomena that are *constructed* out of the natural phenomena which our experience—as it is formed and informed by practical and theoretical ideologies—encounters" (Baltas 1988, pp. 220–21).

Feyerabend has a similar view: "scientists are sculptors of reality—but sculptors in a special sense. They not only *act casually* upon the world (though they do that, too, and must if they want to "discover" new entities); they also create *semantic conditions* that engender strong inferences from known effects to novel projections and, conversely, from the projections to testable effects" (Feyerabend 1990, p. 151).

Latour and Woolgar also conceive facts as constructions. Their view is that "facts are socially constructed" and that "the process of construction involves the use of certain devices whereby all traces of production are made extremely difficult to detect" (Latour and Woolgar 1979, p. 176). The view upheld here is that the construction of facts is only social in the obvious sense that every human activity is, and that what social traces there are in the construction of scientific facts are lost because social factors have to be filtered through the factors of scientific dialectics. If there is any social conditioning in this construction (and there is), it is not direct but mediated by this filter. The same observations made about Bloor's view can be applied here (note 30, chap. 5): what acts directly on knowledge is the action of the factors of scientific dialectics.

21. Hacking (1983), p. 27.

22. Ibid., p. 22.

23. Hacking (1984), p. 167.

24. McMullin (1984), p. 23.

25. Quoted in Harré (1983), p. 164.

26. The expression, and the dread associated with it, comes from Putnam (1978), p. 25.

27. McMullin (1984), p. 22.

28. There are at least three labels. The first is "semantic realism," according to which "claims about theoretical entities should be taken at 'face value'" and "whatever microphysical facts there are need not be discoverable by us and do not depend upon our methodology" (Horwich 1982, p. 182). The second is "minimal realism," or "NOA" (natural ontological attitude), according to which we must "try to take science on its own terms, and try not to read things into science" (Fine 1986, p. 149), which means that we must combine our acceptance of "the results of science as true" with "the stubborn refusal to amplify the concept of truth by providing a theory or analysis (or even a metaphysical picture)" (pp. 130 and 133). The third label is "realism without correspondence," according to which "we can say of terms in scientific propositions that they refer to elements in a human-independent reality, but not that these propositions describe reality as it is 'in itself' (approximately or not)" (Radder 1988, p. 102).

29. Kant takes for granted the existence of the entities of common experience and science. Since he conceives of these entities as real objects and not private sense-data, Kant is an "empirical realist." As he writes (1783, p. 36): "I grant by all means that there are bodies without us, that is, things which, though quite unknown to us to what they are in themselves, we yet know by the representation which their influence on our sensibility procures us. These representations we call 'bodies,' a term signifying merely the appearance of the thing which is unknown to us, but not therefore less actual. Can this be termed idealism? It is the very contrary."

But Kant also recognizes that our experience of these objects is subjected to the conditions of our mind (forms of intuition and categories, but not only these; see note 7 above). Thus, considered transcendentally (as subjected to these conditions), real objects are phenomena and not things in themselves. Hence Kant is also a "transcendental idealist." But transcendental idealism is not the same as idealism because it "concerns not the existence of things (the doubt of which, however, constitutes idealism in the ordinary sense), since it never came into my head, but it concerns the sensuous representation of things to which space and time specifically belong" (1783, p. 41). Similarly, empirical realism is not the same as realism (in the sense of realism of entities or any other version examined in the text), because it claims it is meaningless to say that the sensuous representation of things "is quite similar to the object—an assertion in which I can find as little meaning as if I said that the sensation of red has a similarity to the property of cinnabar which excites this sensation in me" (1783, p. 37). Since Kant here says it is meaningless to say that a representation (concept) corresponds to a property (object), his empirical realism is not based on the correspondence theory of truth.

The distinction between the reality of the objects considered with regard to themselves, and the ideality of the objects considered with regard to the forms of our intuition and understanding, can also be seen in this passage from a letter written by Kant to J. S. Beck (December 4, 1792) in which he again defends himself against charges of idealism: "Messrs. Eberhard's and Garve's opinion that Berkeley's idealism is the same as that of the critical philosophy (which I could better call 'the principle of the ideality of space and time') does not deserve the slightest attention. For I speak of ideality in reference to the *form of representations;* but they interpret this to mean ideality with respect to the *matter,* that is, the ideality of the *object* and its very existence" (Kant 1967, p. 198).

30. This is how Putnam depicted an earlier view of his. See Putnam (1981), p. 49.

31. Unfortunately, the inventor of internal realism is not as precise as would be necessary about the meaning of "conceptual scheme." At times Putnam speaks of "a theory or description" (Putnam 1981, p. 49); at others of "our belief system" (p. 50). The point is very important because the old saying "Birds of a feather flock together" applies here too: tell me which unit you internalize in relation to and I shall tell you what sort of realist (or idealist) you are. The spectrum of options may run from Kant (same categorical system, same world) to Feyerabend (every E- or I-theory its own world).

32. A criterion such as this can probably be gleaned from Nola (1980).

33. Anyone suspicious of introducing biology and physiology into philosophical argumentation can follow a different path and claim, with Wittgenstein, that the question of correctness (and change) of certain empirical propositions cannot be raised because they are the "river-bed" made of "hard rock" of our "frame of reference," or our "picture of the world." See Wittgenstein (1969), #83: "The *truth* of certain empirical propositions belongs to our frame of reference"; and #94: "But I did not get my picture of the world by satisfying myself of its correctness; nor do I have it because I am satisfied of its correctness. No: it is the inherited background against which I distinguish between true and false."

34. Propositions referring to facts such as these possess, perhaps for different reasons, the characteristics of Goodman's statements with "initial credibility" (1952) and Quine's observational sentences. See, for example, Quine (1974), p. 39: "a sentence is observational insofar as its truth value, on any occasion, would be agreed to by just about any member of the speech community witnessing the occasion."

35. This is Lorenz and von Bertalanffy's view. The former wrote that "certainly the basically imperfect and coarse reports on the external world which our apparatus for organizing a world picture give us have their correspondence in properties belonging to the thing-in-itself" (Lorenz 1959, p. 41). The latter has claimed that "the fact that animals and human beings are still in existence, proves that their forms of experience correspond to some degree with reality" (von Bertalanffy 1955, p. 79).

36. Lorenz says that "the fin of the fish has not dictated to water its physical properties, just as the eye does not determine the properties of light" (Lorenz 1959, p. 40).

37. This is also Barone's view. He observed that "theory makers taken as specific scientific hypotheses . . . are all men that, under normal conditions, have a common physiological structure allowing them to observe common facts." See Barone (1981), p. 225.

38. Quine (1953), p. 43.

39. This is the same concept of equivalence that is found in Putnam (1983), vol.3, chap. 2.

40. Feyerabend (1987), p. 272. It is quite possible that equivalent theories in our sense are incommensurable in Feyerabend's sense, but I think further qualification is needed for Feyerabend's view according to which in the transition from T to T' between incommensurables "not a single primitive descriptive term of T can be incorporated into T'. . . . The meanings of all descriptive terms of the two theories, primitive as well as defined terms, will be different" (Feyerabend 1965b, p. 115). In the transition between two equivalent systems many terms stay unchanged; only the *relevant* terms (concepts, objects) change meaning.

41. I examined the history and other philosophical aspects of this controversy in Pera (1988) (1989b) and (1992); see also (1991a).

42. Galvani (1841), p. 322.

43. Cases like that of animal electricity, though rare, are not unique. Take Newton's experiment reported in Section 3, for instance. The common fact A = "colored spectrum projected by the prism" is as ambiguous as the fact "animal electricity." Under Newton's interpretation (I-theory) that rays are original qualities, this fact becomes A_1 = "colors split by the prism," while under Hooke's I-theory that light is a simple substance, it becomes A_2 = "colors formed by the prism." Here too the common fact is irrelevant while the relevant facts are not common. See Mamiani's excellent analysis (1986), chap. 4.

44. Volta (1918), 2:210; quoted in Pera (1992), p. 174.

45. Darwin (1859), p. 469. The narrative (and the rhetorical) function of the first-person singular or plural pronoun in the following and in other revised passages of the *Origin* is rightly stressed by Richards (1992).

46. From an unpublished letter from Lyell to Darwin quoted by Himmelfarb (1959), p. 257. Actually, Lyell's attitude was considerably more problematic.

47. Darwin (1903), 1:241.

48. Darwin knew this perfectly well: "If some four or five *good* men come round nearly to our view, I shall not fear ultimate success," he wrote to Hooker. See Darwin (1887), 2:20.

49. Quine (1975), p. 328.

Seven

1. Galileo (1953), p. 38.

2. See Sarton (1957), p. 5: "In fact, progress has no definite and unquestionable meaning in other fields than the field of science."

3. See Popper (1963), p. 216: "science is one of the very few activities — perhaps the only one — in which errors are systematically criticized and fairly often, in time, corrected. This is why we can say that, in science, we often learn from our mistakes, and why we can speak clearly and sensibly about making progress there." See, in addition, Popper (1970), p. 57: "in science (and only in science) can we say that we have made genuine progress, that we know more than we did before"; and Popper (1975), pp. 94–95: "as a matter of historical fact, the history of science is, by and

large, a history of progress. (Science seems to be the only field of human endeavor of which this can be said)."

4. The view that there is no progress in art has often been contested. Croce wrote that "to conceive the history of the artistic production of the human race as developed along a single line of progress and regress would therefore be altogether erroneous" (Croce 1902, p. 224). However, he made a distinction between two cases. In the first, "when many are at work at the same subject, without succeeding in giving to it a suitable form, yet drawing always more nearly to it, there is said to be progress" (pp. 224–25). In the second, "where the subject-matter is not the same, a progressive cycle does not exist" (p. 226).

5. Laudan (1977), p. 146, has claimed that "it is thus possible . . . to be able to compare the progressiveness of different research traditions, *even if those research traditions are utterly incommensurable in terms of the substantive claims they make about the world,*" because "there are many criteria for the comparison of competing theories which do not require any degree of commensurability at the observational level," for example, consistency, simplicity, accuracy of predictions, etc. This means that two research traditions refer to a common domain. If they did not, as Kordig (1971, p. 118) argued against a radical interpretation of Feyerabend's and Kuhn's idea of incommensurability, they could stand in the same relation to each other as, say, that of quantum mechanics and Keynesian theory. Laudan himself admits that for every pair of research traditions there are problems that are "permanent" (1977, p. 140) and "neutral with respect to the various theories which attempt to solve the problems" (1977, p. 145).

6. See Laudan (1984), pp. 64–66.

7. This model can be derived from Whewell (1857), pp. 7ff.

8. This is Popper's view. See Popper (1959), p. 276: "a theory which has been well corroborated can only be superseded by one of a higher level of universality; that is, by a theory which is better testable and which, in addition, *contains* the old, well corroborated theory." Popper (1975), p. 94: "a new theory, however revolutionary, must always be able to explain fully the success of its predecessors." See also Popper (1972), pp. 52–53, on the relations between Einstein's and Newton's theories. Krajewski introduces a variation: "one initial revolution (without correspondence relation) and many occasional revolutions (with correspondence relation)" (Krajewski, 1977, p. 90).

Krajewski's view was anticipated by Kant in his Preface to the second edition of the *Critique of Pure Reason.* There Kant claims that scientific method was responsible for the "intellectual revolution" (Kant 1978, B xi and xii) in mathematics and physics, and that once the initial revolution has taken place, a discipline is no longer "compelled to retrace its steps and strike into some new line of approach" (B vii).

9. See Niiniluoto (1987b), p. 158: "the step from g to g' is progressive if and only if $Tr(g, h^*) < Tr(g,' h^*)$," where Tr is truthlikeness, g and g' are rival answers to a cognitive problem and h^* is an unknown target. See also Niiniluoto (1979), (1980), (1991), as well as Boyd (1973), Putnam (1978), and Newton-Smith (1981).

10. This is Lakatos's definition of empirical and theoretical progress (1970).

11. This is Laudan's model of progress: "progress can occur if and only if the succession of scientific theories in any domain shows an increasing degree of problem-solving effectiveness" (Laudan 1977, p. 68).

12. With some approximation, this is Kuhn's original model (1962), chap. 13, which underscores the fact that the problem-solving capacity is one of the arguments of supporters of a new paradigm.

13. Laudan (1990b), p. 19; (1991), p. 562.

14. Laudan et al. (1988), p. 40.

15. It could be objected that reference to epistemic values is unnecessary. There is a Baconian idea according to which a theory is progressive with respect to another if it allows greater control over nature. Rescher, for example, claims that "in the first analysis, *praxis* is the arbiter of theory. To understand scientific progress and its limits, we must look not towards the dialectic of questions and answers but towards the scope and limits of human power in our transactions with nature" (Rescher 1984, p. 46), and we must "step outside the domain of pure theory into that of practice" (p. 35). Undeniably, this answer highlights an important point: our current theories allow us to accomplish the unthinkable. We could not have gone to the moon on the basis of Aristotle's theory of movement, and a normal heart bypass operation would not be possible using Galen's biology. And yet the answer does not suffice. It is not enough to say *that* our theories have more practical success than their predecessors; we would also like to know *why,* that is, we would like to know for what epistemic reasons. Without these, a physicist's success at manipulating things in his laboratory is no different from a cook's success at manipulating things in the kitchen. Bacon could infer the epistemic value of theories from their practical success because he claimed that "the fruits and works are as it were sponsors and sureties for the truth of philosophies" (1620, I, 73, p. 73) and that "that which has maximum utility in practice has maximum truth in knowledge" (1620, II, 4). This is, perhaps, an understandable act of faith for a lord chancellor but cannot satisfy a modern philosopher. The Voltaic cell was highly successful in practice, but the theory behind it, upheld by Volta, was erroneous.

16. Kuhn (1962), p. 42.

17. See Kuhn (1970a), p. 262: "what I am denying then is neither the existence of good reasons nor that these reasons are of the sort usually described. I am, however, insisting that such reasons constitute values to be used in making choices rather than rules of choice."

18. See Kuhn (1977), p. 331: "subtract accuracy of fit to nature from the list and the enterprise that results may not resemble science at all, but perhaps philosophy instead."

19. Ibid., p. 325.

20. Ibid.

21. Laudan (1984), p. 92.

22. Ibid., p. 98.

23. Ibid., p. 99; my emphasis.

24. Ibid., p. 92.

25. Ibid.

26. There can nonetheless be problems for the *M-A* pair. Suppose T_1 and T_2 are two research traditions that share "problem-solving effectiveness." As has been seen (Doppelt 1983, p. 133), the point is that, in Laudan's words (1977, p. 133), "an approximate determination of the effectiveness of a research tradition can be made *within* the tradition itself, without reference to any other research tradition." It would seem then that sharing this value is not enough to shift consensus from T_1 to T_2.

27. Laudan (1981), p. 149.

References

Agazzi, E. 1976. "The Concept of Empirical Data." In *Formal Methods in the Methodology of Empirical Sciences,* ed. M. Przelecki, K. Szaniawski, and R. Wojcicki, pp. 143–57. Wroclaw: Ossolineum.

———. 1989. "Naive Realism and Naive Antirealism." *Dialectica* 43:83–98.

Apostel, L. 1979. "What Is the Force of an Argument? Some Problems and Suggestions." *Revue Internationale de Philosophie* 127–28:99–109.

Aristotle. *On Generation and Corruption.* In *The Complete Works of Aristotle.* 2 vols., ed. J. Barnes. Bollingen Series 71. Princeton: Princeton University Press, 1984.

———. *On the Heavens.* In *The Complete Works of Aristotle.*

———. *Posterior Analytics.* In *The Complete Works of Aristotle.*

———. *Prior Analytics.* In *The Complete Works of Aristotle.*

———. *Rhetoric.* In *The Complete Works of Aristotle.*

———. *Sophistical Refutations.* In *The Complete Works of Aristotle.*

———. *Topics.* In *The Complete Works of Aristotle.*

Bacon, F. 1607. *Cogitata et Visa de Interpretatione Naturae.* In *The Works of Francis Bacon.* 4 vols., ed. J. Spedding, R. L. Ellis, and D. D. Heath. London: Longman, 1860.

———. 1620a. *Novum Organum.* In *The Works of Francis Bacon,* vol. 1.

———. 1620b. *The Great Instauration.* In *The Works of Francis Bacon,* vol. 4.

———. 1623. *On the Dignity and Advancement of Learning.* In *The Works of Francis Bacon,* vol. 3.

Baltas, A. 1987. "Ideological 'Assumptions' in Physics: Social Determinations of Internal Structures." In *PSA 1986.* 2 vols. Vol. 2, ed. A. Fine and P. Machamer, pp. 130–51. East Lansing, Mich.: Philosophy of Science Association.

———. 1988. "On the Structure of Physics as a Science." In *Theory and Experiment,* ed. D. Batens and J. P. van Bendegem, pp. 207–25. Dordrecht and Boston: Reidel.

Barker, S. F. 1965. "Must Every Inference Be Either Deductive or Inductive?" In *Philosophy in America,* ed. M. Black, pp. 58–73. Ithaca: Cornell University Press.

Barnes, B., and D. Bloor. 1982. "Relativism, Rationalism, and the Sociology of Knowledge." In *Rationality and Relativism,* ed. M. Hollis and S. Lukes, pp. 21–47. Oxford: Basil Blackwell.

Barnes, J., M. Schofield, and R. Sorabji, eds. 1975. *Articles on Aristotle.* Vol. 1, *Science.* London: Duckworth.

Barone, F. 1967. "La logica in Galileo." In Barone, *Immagini filosofiche della scienza,* pp. 73–95.

———. 1981. "Metodologia e storiografia della scienza del Novecento." In Barone, *Immagini filosofiche della scienza*, pp. 209–27.

———. 1983. *Immagini filosofiche della scienza.* Rome-Bari: Laterza.

Barrotta, P. 1992. *Gli argomenti dell'economia.* Milan: Franco Angeli.

Barth, E. M., and E. C. W. Krabbe. 1982. *From Axiom to Dialogue: A Philosophical Study of Logics and Argumentation.* Berlin and New York: Walter de Gruyter.

Bernard, Cl. [1865] 1927. *An Introduction to the Study of Experimental Medicine.* Trans. H. C. Greene. New York: Macmillan.

Bernstein, R. 1983. *Beyond Objectivity and Relativism.* Philadelphia: University of Pennsylvania Press.

Bertalanffy, L. von. 1955. "An Essay on the Relativity of Categories." Reprinted in *General Systems,* vol. 7, ed. L. von Bertalanffy and A. Rapoport, pp. 71–83.

Berti, E. 1987. *Contraddizione e dialettica negli antichi e nei moderni.* Palermo: L'Epos.

———. 1989. *Le ragioni di Aristotele.* Rome-Bari: Laterza.

Beveridge, W. I. B. 1980. *Seeds of Discovery.* London: Heinemann.

Bloor, D. 1976. *Knowledge and Social Imagery.* 2d ed. 1991. London: Routledge and Kegan Paul.

Bondi, H. 1960a. *The Universe at Large.* London: Heinemann.

———. 1960b. *Cosmology.* 2d ed. Cambridge: Cambridge University Press.

Bondi, H., and T. Gold. 1948. "The Steady-State Theory of the Expanding Universe." *Monthly Notices of the Royal Astronomical Society* 108, no. 3: 252–70.

Bondi, H., B. Bonnor, A. Lyttleton, and G. J. Whitrow. 1960. *Rival Cosmological Theories.* Oxford: Oxford University Press.

Bonitz, H. 1955. *Index Aristotelicus.* 2d. ed. Graz: Akademische Druck-U. Verlagsanstalt.

Boyd, R. 1973. "Realism, Underdetermination, and a Causal Theory of Evidence." *Nous* 7:1–12.

Boyd, R. N. 1984. "The Current Status of Scientific Realism." In *Scientific Realism,* ed. J. Leplin, pp. 41–82. Berkeley: University of California Press.

Bozzi, P. 1990. *Fisica ingenua.* Milan: Garzanti.

Brown, H. I. 1977. *Perception, Theory, and Commitment: The New Philosophy of Science.* Chicago: University of Chicago Press.

———. 1987. *Observation and Objectivity.* Oxford: Oxford University Press.

Bunge, M. 1967. *Scientific Research.* 2 vols. Berlin: Springer-Verlag.

Caplan, A. L. 1979. "Darwinism and Deductivist Models of Theory Structure." *Studies in History and Philosophy of Science* 10:341–53.

Carnap, R. 1959. *Introduction to Semantics and Formalization of Logic.* Cambridge: Harvard University Press.

———. 1962. *Logical Foundations of Probability.* 2d ed. Chicago: University of Chicago Press.

Cartwright, N. 1983. *How the Laws of Physics Lie.* Oxford: Oxford University Press.

Croce, B. [1902] 1909. *Aesthetic as Science of Expression and General Linguistic.* Trans. D. Ainslie. London: Macmillan.

Crombie, A. C. 1960. "Darwin's Scientific Method." In *Actes du IXe Congrès International d'Histoires des Sciences.* Vol. 1: 354–62. Paris: Blanchard.

Darlington, C. D. 1959. *Darwin's Place in History.* Oxford: Basil Blackwell.

Darwin, C. [1859] 1975. *On the Origin of Species,* A Facsimile of the First Edition. Introduction by E. Mayr. Cambridge: Harvard University Press.

———. [1872] 1972. *The Origin of Species.* 6th ed. London: Dent & Sons.

———. 1887. *The Life and Letters of Charles Darwin, including an Autobiographical Chapter.* 2 vols., ed. F. Darwin. Reprinted in *The Works of Charles Darwin,* vols. 17–19. New York: AMS Press, 1972.

———. 1903. *More Letters of Charles Darwin.* 2 vols., ed. F. Darwin and A. C. Seward. London: Murray.

———. 1958. *The Autobiography of Charles Darwin,* ed. N. Barlow. New York: W. W. Norton.

Davidson, D. 1974. "*On the Very Idea of a Conceptual Theme.*" Reprinted in *Inquiries into Truth and Imagination.* Oxford: Clarendon Press.

Davies, P. 1977. *Space and Time in the Modern Universe.* Cambridge: Cambridge University Press.

De Finetti, B. 1972. *Probability, Induction, and Statistics.* London: J. Wiley and Sons.

De Pater, W. A. 1965. *Les topiques d'Aristote et la dialectique platonicienne.* Fribourg: Editions St. Paul.

Descartes, R. [1628] 1985. *Rules for the Direction of the Mind.* In *The Philosophical Writings of Descartes.* 2 vols., trans. J. Cottingham, R. Stoothoff, and D. Murdoch, vol. 1, pp. 9–78. Cambridge: Cambridge University Press.

Deutsch, E., ed. 1991. *Culture and Modernity: East-West Philosophic Perspectives.* Honolulu: University of Hawaii Press.

Doppelt, G. 1978. "Kuhn's Epistemological Relativism: An Interpretation and Defense." *Inquiry* 21:33–86.

———. 1983. "Relativism and Recent Pragmatic Conceptions of Scientific Rationality." In *Scientific Explanation and Understanding,* ed. N. Rescher, pp. 106–42. Lanham, Md.: University Press of America.

Dretske, F. 1969. *Seeing and Knowing.* Chicago: University of Chicago Press.

Eemeren, F. H. van, R. Grootendorst, and T. Kruiger. 1984. *The Study of Argumentation.* New York: Irvington.

Egerton, F. N. 1971. "Darwin's Method or Methods?" *Studies in History and Philosophy of Science* 2:281–86.

Einstein, A., and M. Born. 1971. *The Born-Einstein Letters,* trans. I. Born. New York: Walker.

Eysenck, H. J. 1953. *Uses and Abuses of Psychology.* Harmondsworth: Penguin.

Federspil, G., and C. Scandellari. 1984. "Medicina scientifica e medicina alternativa." *Medicina-Riv. E.M.I.* 4:433–42.

Feyerabend, P. K. 1962. "Explanation, Reduction, and Empiricism." In Feyerabend, *Philosophical Papers,* 1:44–96.

———. 1965a. "Problems of Empiricism." In *Beyond the Edge of Certainty,* ed. R. G. Colodny. Lanham, Md.: University Press of America.

———. 1965b. "Reply to Criticism: Comments on Smart, Sellars, and Putnam." In Feyerabend, *Philosophical Papers,* 1:104–31.

———. 1969. "Science without Experience." In Feyerabend, *Philosophical Papers,* 1:132–35.

———. 1970. "Consolations for the Specialist." In *Criticism and the Growth of Knowledge,* ed. I. Lakatos and A. Musgrave, pp. 197–230. Cambridge: Cambridge University Press.

———. 1975. *Against Method.* London: New Left Books.

———. 1976. "The Methodology of Scientific Research Programmes." In Feyerabend, *Philosophical Papers,* 2:202–30.

———. 1978. *Science in a Free Society.* London: New Left Books.

———. 1981. *Philosophical Papers.* 2 vols. Cambridge: Cambridge University Press.

———. 1987. *Farewell to Reason.* London and New York: Verso.

———. 1990. "Realism and the Historicity of Knowledge." In *Creativity in the Arts and Science,* ed. W. R. Shea and A. Spadafora. Canton, Mass.: Watson Publishing International.

Fine, A. 1986. *The Shaky Game: Einstein, Realism, and the Quantum Theory.* Chicago: University of Chicago Press.

Finocchiaro, M. 1980. *Galileo and the Art of Reasoning.* Dordrecht and Boston: Reidel.

———. 1981. "Fallacies and the Evaluation of Reasoning." *American Philosophical Quarterly* 18:13–22.

———. 1989. *The Galileo Affair: A Documentary History.* Berkeley: University of California Press.

———. 1990. "Varieties of Rhetoric in Science." *History of the Human Sciences* 3:177–93.

Franklin, B. 1941. *Benjamin Franklin's Experiments,* ed. I. B. Cohen. Cambridge: Harvard University Press.

Freud, S. 1922. "Psycho-analysis." In *The Standard Edition of the Complete Psychological Works of Sigmund Freud,* ed. J. Strachey et al., vol. 18, pp. 235–54. London: Hogarth Press, 1953–74.

———. 1932. *New Introductory Lectures on Psycho-analysis.* In *The Standard Edition of the Complete Psychological Works of Sigmund Freud,* 22:3–182.

Gale, B. G. 1982. *Evolution without Evidence: Charles Darwin and the Origin of Species.* Brighton: Harvester Press.

Galilei, G. 1593. *Le mecaniche.* In *Le opere di Galileo Galilei,* 20 vols., national edition, ed. A. Favaro. 2:147–91. Florence: Barbèra.

———. 1605. *Trattato della sfera ovvero cosmografia.* In *Le opere di Galileo Galilei,* 2:211–55.

———. 1610. "Lettera a G. Keplero (19 agosto 1610)." In *Le opere di Galileo Galilei,* vol. 10.

———. 1611. "Lettera a Gallanzone Gallanzoni (18 luglio 1611)." In *Le opere di Galileo Galilei,* 11:141–55.

———. 1613. *Istoria e dimostrazioni matematiche intorno alle macchie solari e loro accidenti,* in *Le opere di Galileo Galilei,* 5:71–249.

———. 1615. "Considerations on the Copernican Opinion." In M. Finocchiaro, *The Galileo Affair: A Documentary History,* pp. 70–86. Berkeley: University of California Press, 1989.

———. 1623. *Il saggiatore.* In *Le opere di Galileo Galilei,* 6:197–372.

———. 1624. "Reply to Ingoli." In Finocchiaro, *The Galileo Affair: A Documentary History,* pp. 154–97. Berkeley: University of California Press, 1989.

———. 1633. "Note" to *Esercitazioni filosofiche di Antonio Rocco.* In *Le opere di Galileo Galilei,* 7:569–750.

———. 1953. *Dialogue Concerning the Two Chief World Systems,* trans. S. Drake. Berkeley and Los Angeles: University of California Press.

———. 1974. *Two New Sciences,* trans. S. Drake. Madison: University of Wisconson Press.

Galvani, L. 1841. *Opere edite e inedite del Professore Luigi Galvani.* Accademia delle Scienze dell'Istituto di Bologna. Bologna: E. Dall'Olmo.

Gassendi, P. 1658a. *Syntagma Philosophicum.* In *Opera Omnia,* vol. 1. Stuttgart and Bad Canstatt: Fromman, 1964.

———. 1658b. *Exercitationes Paradoxicae adversus Aristoteleos.* In *Opera Omnia,* vol. 3.

Ghiselin, M. T. 1969. *The Triumph of the Darwinian Method.* Berkeley: University of California Press.

Giuliani, A. 1986. "Logica del diritto." In *Enciclopedia del diritto.* Milan: Giuffrè.

Goethe, J. W. 1949. *Maximen und Reflexionen.* In J. W. Goethe, *Gedenkausgabe der Werke, Briefe und Gespräche,* vol. 9. Zurich: Artemis Verlags.

Goldberg, S. 1984. *Understanding Relativity.* Boston: Birkhauser.

Goodman, N. 1952. "Sense and Certainty." *Philosophical Review* 61:160–67.

Govier, T. 1987. "Beyond Induction and Deduction." In *Argumentation: Across the Lines of Discipline,* ed. F. H. van Eemeren, R. Grootendorst, J. A. Blair, and C. A. Willard. Dordrecht: Foris Publications.

Gross, A. G. 1990. *The Rhetoric of Science.* Cambridge: Harvard University Press.

Grünbaum, A. 1984. *The Foundations of Psychoanalysis.* Berkeley: University of California Press.

———. 1986. "Précis of *The Foundations of Psychoanalysis:* A Philosophical Critique." *Behavorial and Brain Sciences* 9:217–84.

———. 1989. "The Pseudo-Problem of Creation in Physical Cosmology." *Philosophy of Science* 56:373–94.

Hacking, I. 1982. "Language, Truth, and Reason." In *Rationality and Relativism,* ed. M. Hollis and S. Lukes, pp. 48–66. Oxford: Basil Blackwell.

———. 1983. *Representing and Intervening.* Cambridge: Cambridge University Press.

———. 1984. "Experimentation and Scientific Reasoning." In *Scientific Realism,* ed. J. Leplin, pp. 154–72. Berkeley: University of California Press.

Hamblin, C. L. 1970. *Fallacies.* London: Methuen.

Hanson, R. N. 1958. *Patterns of Discovery.* Cambridge: Cambridge University Press.

———. 1969. *Perception and Discovery,* ed. W. C. Humphreys. San Francisco: Freeman Cooper.

Harré, R. 1983. *Great Scientific Experiments.* Oxford: Oxford University Press.

Harris, E. 1970. *Hypothesis and Perception.* London: Allen and Unwin.

Heisenberg, W. 1958. *Physics and Philosophy.* New York: Harper and Row.

Hempel, C. 1983. "Valuation and Objectivity in Science." In *Physics, Philosophy, and Psychoanalysis,* ed. R. S. Cohen and L. Laudan, pp. 73–100. Dordrecht and Boston: Reidel.

Herschel, J. F. W. [1830] 1966. *A Preliminary Discourse on the Study of Natural Philosophy.* New York: Johnson Reprint.

Hesse, M. 1980. *Revolutions and Reconstructions in the Philosophy of Science.* Brighton: Harvester Press.

Himmelfarb, G. 1959. *Darwin and the Darwinian Revolution.* 2d ed. New York: Norton Library.

Hintikka, J. 1987. "The Interrogative Approach to Inquiry and Probabilistic Inference." *Erkenntnis* 26:429–42.

———. 1988. "What Is the Logic of Experimental Inquiry?" *Synthese* 74:173–90.

———. 1989. "The Role of Logic in Argumentation." *Monist* 72:3–24.

Hintikka, J., and M. Hintikka. 1982. "Sherlock Holmes Confronts Modern Logic: Toward a Theory of Information-Seeking through Questioning." In *Argumentation: Approaches to Theory Formation,* ed. E. M. Barth and J. L. Martens, pp. 55–76.

Hobbes, T. 1642. *De Cive.* In *The Philosophical Works of Thomas Hobbes,* vol. 3, trans. H. Warrander. Oxford: Clarendon Press.

Hodge, M. J. S. 1977. "The Structure and Strategy of Darwin's 'Long Argument.'" *British Journal for the History of Science* 10:237–45.

Hollis, M., and S. Lukes, eds. 1982. *Rationality and Relativism.* Oxford: Basil Blackwell.

Hooke, R. 1665. *Micrographia.* New York: Dover, 1961.

Hopkins, W. 1860. "Physical Theories of the Phenomena of Life." In D. L. Hull, *Darwin and His Critics,* pp. 229–75. Chicago and London: University of Chicago Press, 1973.

Horwich, P. 1982. "Three Forms of Realism." *Synthese* 51:181–201.

———. 1990. *Truth.* London: Basil Blackwell.

Horwich, P., ed. 1993. *World Changes: Thomas Kuhn and the Nature of Science.* Cambridge: MIT Press.

Hoyle, F. 1948. "A New Model for the Expanding Universe." *Monthly Notices of the Royal Astronomical Society* 108, no. 5: 372–82.

———. 1965. *Galaxies, Nuclei, and Quasars.* New York: Harper and Row.

Hull, D. L. 1973. *Darwin and His Critics.* Chicago and London: University of Chicago Press.

Ingoli, F. 1616. *De situ et quiete terrae contra Copernici systemate disputatio.* In G. Galilei, *Le opere di Galileo Galilei,* 5:403–12.

Jenkin, F. 1867. "The Origin of Species." In D. L. Hull, *Darwin and His Critics,* 302–50. Chicago: University of Chicago Press, 1973.

John, L. ed. 1973. *Cosmology Now.* London: BBC.

Jung, C. G. 1934. "The Relations between the Ego and the Unconscious." In *Collected Works,* 20 vols., ed. H. Read, M. Fordham, and G. Adler, vol. 7, pp. 119–239. New York: Pantheon Books, 1953.

Kalinowski, G. 1965. *Introduction à la logique juridique.* Paris: R. Pichon and R. Durand-Auzias.

Kant, I. 1783. *Prolegomena to Any Future Metaphysics,* ed. L. W. Beck. Indianapolis: Bobbs-Merrill, 1950.

———. 1785. *Groundwork of the Metaphysic of Morals,* ed. H. J. Paton. New York: Harper Torchbooks, 1964.

———. 1786. *Metaphysical Foundations of Natural Science,* trans. J. Ellington. Indianapolis: Bobbs-Merrill, 1970.

———. 1790. *Critique of Judgement,* ed. J. C. Meredith. Oxford: Clarendon Press, 1952.

———. 1967. *Philosophical Correspondence,* ed. A. Zweig. Chicago: University of Chicago Press.

———. 1978. *Critique of Pure Reason,* ed. N. Kemp Smith. London: Macmillan.

———. 1988. *Logic,* trans. R. S. Hartman and W. Schwarz. New York: Dover.

Kavaloski, V. C. 1974. "The 'Vera Causa' Principle: A Historical Study of a Methodological Concept from Newton through Darwin." Ph.D. dissertation, University of Chicago.

Kordig, C. R. 1971. *The Justification of Scientific Change.* Dordrecht and Boston: Reidel.

Krabbe, E. C. W. 1978. "The Adequacy of Material Dialogue-Games." *Notre Dame Journal of Formal Logic* 19:321–30.

———. 1982. "Theory of Argumentation and the Dialectical Garb of Formal Logic." In *Argumentation: Approaches to Theory Formation,* ed. E. M. Barth and J. L. Martens, pp. 123–31. Amsterdam: John Benjamins B. V.

Krajewski, W. 1977. *Correspondence Principle and Growth of Science.* Dordrecht and Boston: Reidel.

Kuhn, T. 1962. *The Structure of Scientific Revolutions.* 2d ed. Chicago: University of Chicago Press, 1970.

———. 1970a. "Reflections on My Critics." In Lakatos and Musgrave, eds., *Criticism and the Growth of Knowledge,* pp. 231–78. Cambridge: Cambridge University Press.

———. 1970b. "Postscript 1969." In Kuhn, *The Structure of Scientific Revolutions,* 2d ed., pp. 174–210.

———. 1977. *The Essential Tension.* Chicago: University of Chicago Press.

Lakatos, I. 1970. "Falsification and the Methodology of Scientific Research Programmes." In Lakatos, *Philosophical Papers,* 2 vols., ed. J. Worrall and G. Currie, vol. 1, pp. 8–101. Cambridge: Cambridge University Press, 1978.

———. 1971. "History of Science and Its Rational Reconstructions." In Lakatos, *Philosophical Papers,* vol. 1, pp. 102–38.

———. 1974. "The Role of Crucial Experiments in Science." *Studies in History and Philosophy of Science* 54:309–25.

———. 1976. "Why Did Copernicus's Research Programme Supersede Ptolemy's?" In Lakatos, *Philosophical Papers,* vol. 1, pp. 168–92.

———. 1978a. *Philosophical Papers.* 2 vols. Ed. J. Worrall and G. Currie. Cambridge: Cambridge University Press.

———. 1978b. "What Does a Mathematical Proof Prove?" In Lakatos, *Philosophical Papers,* vol. 2, pp. 61–69.

Lakatos, I., and A. Musgrave, eds. 1970. *Criticism and the Growth of Knowledge.* Cambridge: Cambridge University Press.

Latour, B., and S. Woolgar. 1976. *Laboratory Life: The Construction of Scientific Facts.* 2d ed. Princeton: Princeton University Press, 1986.

Laudan, L. 1977. *Progress and Its Problems.* Berkeley: University of California Press.

———. 1978. "The Philosophy of Progress . . ." In *PSA 1978,* 2 vols., ed. P. D. Asquith and I. Hacking, vol. 2, pp. 530–47. East Lansing, Mich.: Philosophy of Science Association.

———. 1981. *Science and Hypothesis.* Dordrecht and Boston: Reidel.

———. 1983. "The Demise of the Demarcation Problem." In *Physics, Philosophy, and Psychoanalysis,* ed. R. S. Cohen and L. Laudan, pp. 111–27. Dordrecht and Boston: Reidel.

———. 1984. *Science and Values.* Berkeley: University of California Press.

———. 1987a. "Progress or Rationality? The Prospects for Normative Naturalism." *American Philosophical Quarterly* 24, no. 1: 19–31.

———. 1987b. "Methodology's Prospects." In *PSA 1986,* 2 vols., ed. A. Fine and P. Machamer, vol. 2, pp. 347–54. East Lansing, Mich.: Philosophy of Science Association.

———. 1990a. "Normative Naturalism." *Philosophy of Science* 57:44–59.

———. 1990b. *Science and Relativism.* Chicago: University of Chicago Press.

———. 1991. "Scientific Progress and Content Loss." In E. Deutsch, ed., *Culture and Modernity: East-West Philosophic Perspectives,* pp. 561–69. Honolulu: University of Hawaii Press.

Laudan, L., et al. 1986. "Scientific Change: Philosophical Models and Historical Research." *Synthèse* 69:141–223.

Laudan, R., L. Laudan, and A. Donovan. 1988. "Testing Theories of Scientific Change." In *Scrutinizing Science,* ed. A. Donovan, L. Laudan, and R. Laudan, pp. 3–44. Dordrecht and Boston: Kluwer Academic Press.

Legrenzi, P. 1975. "Introspezione." *Psicologia contemporanea* 11:53–54.

Leibniz, G. W. 1961. *Philosophische Schriften.* 7 vols., herausgegeben von C. I. Gerhardt. Hildesheim: G. Olms.

Leplin, J. 1984. "Introduction." In Leplin, ed., *Scientific Realism,* pp. 1–7. Berkeley: University of California Press.

Lloyd, E. A. 1983. "The Nature of Darwin's Support for the Theory of Natural Selection." *Philosophy of Science* 50:112–29.

Locke, J. 1692. *Essay on Human Understanding.* 2 vols. New York: Dover, 1959.

Lorenz, K. 1941. "Kant's Doctrine of the A Priori in the Light of Contemporary Biology." Reprinted in *General Systems,* vol. 7, ed. L. von Bertalanffy and A. Rapoport (1962), pp. 23–25.

———. 1959. "Gestalt Perception as Fundamental to Scientific Knowledge." Reprinted in *General Systems,* vol. 7 (1962), pp. 37–56.

Lorenzen, P. 1961. "Ein dialogisches Konstruktivitätskriterium." In *Infinitistic Methods: Proceedings of the Symposium on Foundations of Mathematics (Warsaw, 2–9 September 1959),* pp. 193–200. Oxford: Pergamon Press.

Mamiani, M. 1986. *Il prisma di Newton.* Rome-Bari: Laterza.

Mansion, S. 1961. "Le role de l'exposé et de la critique des philosophes antérieures chez Aristote." In *Aristote et les problèmes de méthode.* Louvain: Publication Universitaires de Louvain.

Marconi, D. 1987. *L'eredità di Wittgenstein.* Rome-Bari: Laterza.

Mayr, E. 1991. *Charles Darwin and the Genesis of Modern Evolutionary Thought.* Cambridge: Cambridge University Press.

McCloskey, M. 1983. "Intuitive Physics." *Scientific American* 248, no. 4: 14–22.

McMullin, E. 1979. "The Ambiguity of 'Historicism.'" In *Current Research in Philosophy of Science,* ed. P. D. Asquith and H. E. Kyburg, Jr., pp. 55–83. East Lansing, Mich.: Philosophy of Science Association.

———. 1981a. "How Should Cosmology Relate to Theology?" In *The Sciences and Theology in the Twentieth Century,* ed. A. R. Peacocke, pp. 17–57. Stocksfield: Oriel Press.

———. 1981b. "Is Philosophy Relevant to Cosmology?" *American Philosophical Quarterly* 18:177–89.

———. 1984. "A Case for Scientific Realism." In J. Leplin, ed., *Scientific Realism,* pp. 8–40. Berkeley: University of California Press.

Melia, T. 1992. "Essay Review on P. Dear, *The Literary Structure of Scientific Argument: Historical Studies;* A. G. Gross, *The Rhetoric of Science;* G. Meyers, *Writing Biology: Texts in the Social Construction of Scientific Knowledge;* and L. J. Prelli, *A Rhetoric of Science: Inventing Scientific Discourse.*" *Isis* 83:100–106.

Mill, J. S. 1865. *A System of Logic: Ratiocinative and Inductive.* 9th ed. London: Longmans.

Moss, J. D. 1983. "Galileo's *Letter to Christina:* Some Rhetorical Considerations." *Renaissance Quarterly* 36:547–76.

———. 1984. "Galileo's Rhetorical Strategies in Defence of Copernicanism." In *Novità celesti e crisi del sapere,* ed. P. Galluzzi, 95–103. Florence: Giunti Barbèra.

———. 1986. "The Rhetoric of Proof in Galileo's Writings on the Copernican System." In *Reinterpreting Galileo,* ed. W. W. Wallace. Washington, D.C.: Catholic University of America Press.

———. 1993. *Novelities in the Heavens: Rhetoric and Science in the Copernican Controversy.* Chicago: University of Chicago Press.

Musgrave, A. 1976. "Method or Madness?" In *Essays in Memory of Imre Lakatos,* ed. R. S. Cohen et al., pp. 457–91. Dordrecht and Boston: Reidel.

Narlikar, J. 1973. "Steady-State Defended." In L. John, ed., *Cosmology Now,* pp. 69–84. London: BBC.

Newton, I. 1726. *Philosophiae Naturalis Principia Mathematica,* trans. A. Motte and F. Cajori. Berkeley: University of California Press.

Newton-Smith, W. 1981. *The Rationality of Science.* London: Routledge and Kegan Paul.

Niiniluoto, I. 1979. "Verisimilitude, Theory-Change, and Scientific Progress." In *The Logic and Epistemology of Scientific Change,* ed. I. Niiniluoto and R. Tuomela. Acta Philosophica Fennica 30:243–64.

———. 1980. "Scientific Progress." *Synthèse* 45:427–64.

———. 1987a. "Varieties of Realism." In *Symposium on the Foundations of Modern Physics, 1987,* ed. P. Lahti and P. Millestaedt, pp. 459–83. Singapore: World Scientific.

———. 1987b. "Progress, Realism, and Verisimilitude." In *Logik, Wissenschafttheorie und Erkenntnistheorie.* Akten des 11. Internationalen Wittgenstein Symposiums, pp. 151–61. Vienna: Hölder-Pichler-Tempsky.

———. 1991. "Scientific Progress Reconsidered." In E. Deutsch, ed., *Culture and Modernity: East-West Philosophic Perspectives.* Honolulu: University of Hawaii Press.

Nola, R. 1980. "Fixing the Reference of Theoretical Terms." *Philosophy of Science* 47:505–31.

———. 1986. "Observation and Growth in Scientific Knowledge." In *PSA 1986,* 2 vols., ed. A. Fine and P. Machamer, vol. 1, pp. 245–57. East Lansing, Mich.: Philosophy of Science Association.

Owen, G. E. L. [1961] 1975. "'Tithenai ta Phainomena.'" In J. Barnes, M. Schofield, and R. Sorabji, eds., *Articles on Aristotle,* vol. 1, *Science*, pp. 113–26. London: Duckworth.

Pasquinelli, A. 1968. *Letture galileiane.* Bologna: Il Mulino.

Peirce, C. S. 1877. "The Fixation of Belief." In *The Collected Papers of Charles Sanders Peirce,* ed. C. Hartshorne and P. Weiss, vol. 5, pp. 385–87. Cambridge: Cambridge University Press, 1934.

Pera, M. 1980. "Inductive Method and Scientific Discovery." In *On Scientific Discovery,* ed. M. d. Grmek, R. Cohen, and G. Cimino. Dordrecht and Boston: Reidel.

———. 1982. *Apologia del metodo.* Rome-Bari: Laterza.

———. 1986. "Narcissus at the Pool: Scientific Method and the History of Science." *Organon* 22–23 (1986–87): 79–98.

———. 1987. "The Rationality of Discovery: Galvani's Animal Electricity." In *Rational Changes in Science,* ed. M. Pera and J. Pitt, pp. 177–201. Dordrecht and Boston: Reidel.

———. 1988. "Radical Theory Change and Empirical Equivalence: The Galvani-

Volta Controversy." In *Scientific Revolutions,* ed. W. Shea. Canton, Mass.: Watson Publishing International.

———. 1989a. "Methodological Sophisticationism: A Degenerating Project." In *Imre Lakatos and Theories of Scientific Change,* ed. K. Gavroglu, Y. Goudaroulis, and P. Nicolacopoulos, pp. 160–97. Dordrecht: Kluwer Academic Publishers.

———. 1989b. "How Crucial Is a Crucial Experiment? Reflections on the Galvani-Volta Controversy." In *From Luigi Galvani to Contemporary Neurobiology,* ed. A. Baruzzi et al. Fidia Research Series, vol. 22, pp. 19–37. Padua: Liviana Press; Berlin: Springer Verlag.

———. 1991a. "The Role and Value of Rhetoric in Science." In *Persuading Science: The Art of Scientific Rhetoric,* ed. M. Pera and W. Shea. Canton, Mass.: Watson Publishing International.

———. 1991b. "A Dialectical View of Scientific Rationality and Progress." In E. Deutsch, ed., *Culture and Modernity: East-West Philosophic Perspectives,* pp. 570–92. Honolulu: University of Hawaii Press.

———. 1992. *The Ambiguous Frog: The Galvani-Volta Controversy on Animal Electricity,* trans. J. Mandelbaum (Italian ed. 1986). Princeton: Princeton University Press.

Perelman, Ch. 1963. *The Idea of Justice and the Problem of Argument,* trans. J. Petrie. London: Routledge and Kegan Paul.

———. 1968a. "Le raisonnement pratique." In *Contemporary Philosophy: A Survey,* ed. R. Klibansky, pp. 168–76. Florence: La Nuova Italia Editrice.

———. 1968b. "Recherches interdisciplinaires sur l'argumentation." *Logique et analyse* 11 no. 44: 502–11.

———. 1970. *Le champ de l'argumentation.* Brussels: Presses Universitaires de Bruxelles.

———. 1971. "Law, Logic, and Epistemology." In Perelman, *Justice, Law, and Argument,* pp. 136–47. Dordrecht and Boston: Reidel, 1980.

———. 1977. *The Realm of Rhetoric,* trans. W. Klubach. Notre Dame: University of Notre Dame Press, 1982.

———. 1979. *Logique juridique: Nouvelle rhétorique.* 2d ed. Toulouse: Dalloz.

———. 1980. *Justice, Law, and Argument.* Dordrecht and Boston: Reidel.

Perelman, Ch., ed. 1965. *Les antinomies en droit.* Brussels: Etablissements Emile Bruylant.

———, ed. 1968. *Le problème des lacunes en droit.* Brussels: Etablissements Emile Bruylant.

Perelman, Ch., and L. Olbrechts-Tyteca. 1958. *The New Rhetoric: A Treatise on Argumentation,* trans. J. Wilkinson and P. Weaver. Notre Dame: University of Notre Dame Press, 1969.

Piaget, J. 1956. *Insights and Illusions of Philosophy,* trans. W. Mays. New York and Cleveland: World, 1971.

Planck, M. 1932. "Causation and Free Will: The Answer of Science." In M. Planck, *Where Is Science Going?* New York: W. W. Norton.

Plato. *Phaedrus,* trans. R. Hackfort. In *The Collected Dialogues,* ed. E. Hamilton and H. Cairns. Bollingen Series 71. Princeton: Princeton University Press, 1987.

Polya, G. 1954. *Mathematics and Plausible Reasoning.* 2 vols. 2d ed. Princeton: Princeton University Press, 1968.

Popper, K. 1945. *The Open Society and Its Enemies.* 2 vols. 5th ed. London: Routledge and Kegan Paul, 1966.

———. 1959. *The Logic of Scientific Discovery.* London: Hutchinson.

———. 1963. *Conjectures and Refutations.* London: Routledge and Kegan Paul.

———. 1970. "Normal Science and Its Dangers." In Lakatos and Musgrave, eds., *Criticism and the Growth of Knowledge,* pp. 51–58. Cambridge: Cambridge University Press.

———. 1972. *Objective Knowledge: An Evolutionary Approach.* Oxford: Clarendon Press.

———. 1974. "Replies to My Critics." In *The Philosophy of Karl Popper,* ed. P. A. Schilpp, pp. 961–1197. La Salle, Ill.: Open Court.

———. 1975. "The Rationality of Scientific Revolutions." Reprinted in *Scientific Revolutions,* ed. I. Hacking, pp. 80–106. Oxford: Oxford University Press, 1981.

———. 1983. *Realism and the Aim of Science,* ed. W. W. Bartley III. London: Hutchinson.

———. 1987. "The Myth of the Framework." In *Rational Changes in Science,* ed. Pitt and Pera, pp. 35–62. Dordrecht and Bostón: Reidel.

Poznanski, E. 1968. "Discussion of McKeon's Paper." *Logique et analyse* 41–42:72–73.

Preti, G. 1968. *Retorica e logica.* Turin: Einaudi.

Purtill, R. L. 1967. "Kuhn on Scientific Revolutions." *Philosophy of Science* 34:53–58.

Putnam, H. 1978. *Meaning and the Moral Sciences.* London: Routledge and Kegan Paul.

———. 1981. *Reason, Truth, and History.* Cambridge: Cambridge University Press.

———. 1983. *Philosophical Papers.* 3 vols. Cambridge: Cambridge University Press.

Quine, W. V. O. 1953. *From a Logical Point of View.* 2d ed. Cambridge: Harvard University Press, 1961.

———. 1960. *Word and Object.* Cambridge: MIT Press.

———. 1974. *The Roots of Reference.* La Salle, Ill.: Open Court.

———. 1975. "On Empirically Equivalent Systems of the World." *Erkenntnis* 9:313–28.

Quinn, P. 1971. "Methodological Appraisal and Heuristic Advice: Problems in the Methodology of Scientific Research Programmes." *Studies in History and Philosophy of Science* 3:135–49.

Radder, H. 1988. *The Material Realization of Science.* Assen: Van Gorcum.

Recker, D. A. 1987. "Causal Efficacy: The Structure of Darwin's Argument Strategy in the *Origin of Species.*" *Philosophy of Science* 54:147–75.

Rescher, N. 1977. *Dialectics.* Albany: State University of New York Press.

———. 1984. *The Limits of Science.* Berkeley: University of California Press.

Resnick, R. 1968. *Introduction to Special Relativity.* New York: John Wiley and Sons.

Richards, R. J. 1992. "The Structure of Narrative Explanation in History and Biology." In *History and Evolution,* ed. M. H. Nitecki and D. V. Nitecki. New York: State University of New York Press, pp. 18–53.

Risch, H. 1980. *Corso di agopuntura.* Bologna: Monduzzi Editore.

Ronchi, V. 1978. "Processo alla scienza." *Atti della Fondazione Giorgio Ronchi* 33, no. 6: 835–67.

———. [1939] 1983. *Storia della luce.* Rome-Bari: Laterza.

Rorty, R. 1980. *Philosophy and the Mirror of Knowledge.* Princeton: Princeton University Press.

———. 1982. *Consequences of Pragmatism.* Brighton: Harvester Press.

———. 1985. "Solidarity or Objectivity?" In *Post-Analytic Philosophy,* ed. J. Raichman and C. West, pp. 3–19. New York: Columbia University Press.

———. 1989. *Contingency, Irony, and Solidarity.* Cambridge: Cambridge University Press.

Ruse, M. 1971. "Natural Selection in the *Origin of Species.*" *Studies in History and Philosophy of Science* 4:311–51.

———. 1975a. "Darwin's Theory of Evolution: An Analysis." *Journal of the History of Biology* 8:219–41.

———. 1975b. "Darwin's Debt to Philosophy: An Examination of the Influence of the Philosophical Ideas of John F. W. Herschel and William Whewell on the Development of Charles Darwin's Theory of Evolution." *Studies in History and Philosophy of Science* 2:159–81.

———. 1979. *The Darwinian Revolution.* Chicago: University of Chicago Press.

Salmon, M. H. 1984. *Logic and Critical Thinking.* San Diego: Harcourt Brace Jovanovich.

Salmon, W. 1966. *The Foundations of Scientific Inference.* Pittsburgh: University of Pittsburgh Press.

———. 1984. *Logic.* 3d ed. Englewood Cliffs, N.J.: Prentice-Hall.

———. 1991. "Rationality and Objectivity in Science, *or* Tom Kuhn Meets Tom Bayes." In *Scientific Theories,* ed. C. W. Savage. Minnesota Studies in the Philosophy of Science, vol. 15. Minneapolis: University of Minnesota Press.

Sapir, E. 1929. "The Status of Linguistics as a Science." *Language* 5:207–14.

Sarkar, H. 1983. *A Theory of Method.* Berkeley: University of California Press.

Sarton, G. 1957. *The Study of the History of Science.* Cambridge: Harvard University Press, 1936; New York: Dover, 1957.

Scarpelli, U. 1982. *L'etica senza verità.* Bologna: Il Mulino.

———. 1986. "Gli orizzonti della giustificazione." In *Etica e diritto,* ed. E. Lecaldano, pp. 3–41. Rome-Bari: Laterza.

Scheffler, I. 1967. *Science and Subjectivity.* Indianapolis: Bobbs-Merrill.

Scheiner, C. 1612a. *Tres epistulae de maculis solaribus.* In Galilei, *Le opere di Galileo Galilei,* vol. 5, pp. 25–33.

———. 1612b. *De maculis solaribus et stellis circa Iovem errantibus accuratior disquisitio.* In Galilei, *Le opere di Galileo Galilei,* vol. 5, pp. 39–70.

Schouls, P. A. 1980. *The Imposition of Method: A Study of Descartes and Locke.* Oxford: Clarendon Press.

Sciama, D. 1973. "Cosmological Models." In L. John, ed., *Cosmology Now,* pp. 55–68. London: BBC.

Sedgwick, A. 1860. "Objections to Mr. Darwin's Theory of the Origin of Species." In D. L. Hull, *Darwin and His Critics,* pp. 159–70. Chicago and London: University of Chicago Press, 1973.

Shapere, D. 1966. "Meaning and Scientific Change." In Shapere, *Reason and the Search for Knowledge,* pp. 58–101. Dordrecht and Boston: Reidel, 1984.

———. 1971. "The Paradigm Concept." In Shapere, *Reason and the Search for Knowledge,* pp. 49–57.

———. 1982. "The Concept of Observation in Science and Philosophy." *Philosophy of Science* 49:485–525.

———. 1984. *Reason and the Search for Knowledge.* Dordrecht and Boston: Reidel.

Shea, W. 1972. *Galileo's Intellectual Revolution.* London: Macmillan.

Sklar, H. 1983. *A Theory of Method.* Berkeley: University of California Press.

Sprat, T. 1667. *The History of the Royal Society.* London: Routledge and Kegan Paul, 1959.

Thagard, P. 1978. "The Best Explanation: Criteria for Theory Choice." *Journal of Philosophy* 75:76–92.

Toulmin, S. 1953. *The Philosophy of Science.* London: Hutchinson.

Vickers, B. 1983. "Epideictic Rhetoric in Galileo's *Dialogo.*" *Annali dell'Istituto e Museo di Storia della Scienza di Firenze* 8, no. 2: 69–102.

Volta, A. 1918. *Le opere di Alessandro Volta.* 7 vols. Edizione nazionale. Milan: Hoepli.

Waismann, F. 1968. *How I See Philosophy,* ed. R. Harré. London: Macmillan.

Walton, D. N. 1985. *Arguer's Position: A Pragmatic Study of Ad Hominem Attack, Criticism, Refutation, and Fallacy.* Westport, Conn.: Greenwood Press.

Watkins, J. 1984. *Science and Scepticism.* Princeton: Princeton University Press.

Weil, E. 1951. "The Place of Logic in Aristotle's Thought." In J. Barnes, M. Schofield, and R. Sorabji, eds., *Articles on Aristotle*, vol. 1, *Science*, pp. 88–112. London: Duckworth, 1975.

Weinberg, S. 1977. *The First Three Minutes*. New York: Basic Books.

Wellman, C. 1971. *Challenge and Response: Justification in Ethics*. Carbondale: Southern Illinois University Press.

Whewell, W. 1833. *Astronomy and General Physics*. Bridgewater Treatise no. 3. London: Pickering.

———. 1847. *The Philosophy of the Inductive Sciences*. 2d ed. New York: Johnson Reprint Corp., 1967.

———. 1857. *History of the Inductive Sciences*. 3d ed., ed. J. W. Parker. Reprint. Hildesheim and New York: Georg Olms Verlag, 1976.

Whorf, B. L. 1954. *Language, Thought, and Reality*. Cambridge: MIT Press.

Winch, P. 1958. *The Idea of Social Science*. London: Routledge and Kegan Paul.

Wittgenstein, L. 1969. *On Certainty*. New York: Harper and Row.

Woods, J., and D. Walton. 1982. *Argument: The Logic of the Fallacies*. Toronto: McGraw-Hill Ryerson.

Index